Quantum Mechanics and Experience

D1331499

Quantum Mechanics and Experience

· · · · · ·

David Z Albert

Harvard University Press

· · ·

Cambridge, Massachusetts
London, England

This book has been digitally reprinted. The content remains
identical to that of previous printings.

Excerpt from "Mr. Cogito and the Imagination" on page xi © 1985 by Zbigniew
Herbert, translation © 1985 by John Carpenter and Bogdana Carpenter. From
Report from the Besieged City and Other Poems by Zbigniew Herbert. First
published by The Ecco Press. Reprinted by permission.

First Harvard University Press paperback edition, 1994

Library of Congress Cataloging-in-Publication Data
Albert, David Z
    Quantum mechanics and experience / David Z Albert.
        p.   cm.
    Includes bibliographical references and index.
    ISBN 0-674-74112-9 (alk. paper) (cloth)
    ISBN 0-674-74113-7 (pbk.)
    1. Physical measurements. 2. Quantum theory. I. Title.
    QC39.A4 1992
    530.1'6—dc20   92-30585
                                                        CIP

To my wife, Orna

Contents

Preface

This book was written both as an elementary text and as an attempt to add to what we presently understand, at the most advanced level, about what seems to me to be the central difficulty at the foundations of quantum mechanics, which is the difficulty about *measurement*.

The first four chapters are a more or less straightforward introduction to that difficulty: Chapter 1 is about the idea of superposition, which is what most importantly distinguishes the quantum-mechanical picture of the world from the classical one, and which is where everything that's puzzling about quantum mechanics comes from. Chapter 2 sets up (in a way that presumes nothing at all, insofar as I understand how to do that, about the mathematical preparation of the reader) the standard quantum-mechanical formalism and outlines the conventional wisdom about how one ought to *think about* that formalism. Chapter 3 is about the Einstein-Podolsky-Rosen argument and how that argument was stunningly undercut by Bell (and it is urged there, by the way, that what Bell's discovery actually amounts to is very frequently misunderstood; it is urged that Bell discovered something not merely about hidden-variable theories but also about *quantum mechanics,* and also about *the world*). Finally, Chapter 4 explicitly sets up the measurement problem.

The rest of the book (which is the *bulk* of the book) is taken up with investigations of those ideas about what to *do* about the measurement problem which seem to me to have some possibility of being right: Chapter 5 is an account and a critique of the idea of the collapse of the wave function (with a detailed discussion of the recent breakthrough of Ghirardi, Rimini, and Weber). Chapter 6 is about a certain very confused but nonetheless (I want to argue) very interesting tradition of thinking about the measurement problem which is (misleadingly) called the "many-worlds" interpretation of quantum mechanics. Chapter 7 is about a completely deterministic replacement for quantum mechanics due to de Broglie and Bohm and Bell. And Chapter 8 is about what the mental lives of sentient observers can potentially be like, if either one of the proposals discussed in Chapters 6 and 7 should actually happen to pan out.

Lots of people helped me out with this. Let me mention a few.

Barry Loewer is the one who first suggested that this book be written, and he has (astonishingly) been willing to spend many hours of his time talking about it with me, and many of the original ideas in it are (as the reader will learn from the references) partly his; and if not for all that, it simply could not have come into being.

I've learned a great deal about the foundations of quantum mechanics from innumerable conversations, over many years, with (first and foremost) Yakir Aharonov, and also with Hilary Putnam, David Deutsch, Irad Kimchie, Marc Albert, Gary Feinberg, Lev Vaidman, Sidney Morgenbesser, Isaac Levi, Shaughn Lavine, and Jeff Barrett, and also with students in some classes I've taught.

I am much indebted to Andrea Kantrowitz for doing such a great job with the illustrations; and I am thankful to Lindsay Waters and Alison Kent and especially Kate Schmit of Harvard University Press, without whose help and understanding this would have been a much less valuable book.

And maybe it ought to be mentioned that this book was written in the hope of finally being able to explain these matters to the reasonable satisfaction of my uncle, the physicist Arthur Kantrowitz, who first got me interested in science.

a bird is a bird
slavery means slavery
a knife is a knife
death remains death

—*Zbigniew Herbert*

Superposition

Here's an unsettling story (the *most* unsettling story, perhaps, to have emerged from any of the physical sciences since the seventeenth century) about something that can happen to electrons. The story is true. The experiments I will describe have all actually been performed.[1]

The story concerns two particular physical properties of electrons which it happens to be possible to measure (with currently available technology) with very great accuracy. The precise physical definitions of those two properties don't matter. Let's call one of them the "color" of the electron, and let's call the other one its "hardness."[2]

It happens to be an empirical fact that the color property of electrons can assume one of only two possible values. Every electron which has thus far been encountered in the world has been either a black electron or a white electron. None have ever been

1. As a matter of fact, not all of these experiments have actually been carried out on *electrons;* in some cases the particles involved were neutrons, and in other cases the "particles" were atoms of silver. Nonetheless, all the experiments described in this story have actually been carried out on one sort of particle or another; and as the reader shall see, the identities of the particles will turn out to be completely irrelevant to our concerns here.

2. One of the properties I have in mind here is (as it happens) the angular momentum with which the electron is spinning about an axis which passes through its center and which runs along the x-direction, and the other one is the angular momentum with which the electron is spinning about an axis which passes through its center and which runs along the y-direction. But all that (as I said) doesn't matter. There are lots of different measurable properties of physical systems that could serve our purposes here just as well.

found to be blue or green. The same goes for hardness. All electrons are either soft ones or hard ones. No one has ever seen an electron whose hardness value was anything other than one of those two.

It's possible to build something called a "color box," which is a device for measuring the color of an electron and which works like this: The box (see figure 1.1) has three apertures. Electrons are fed into the box through the aperture on the left, and every black electron fed in through that aperture exits (along the indicated dashed line) through the aperture marked *b*, and every white electron fed in through that aperture on the left exits through the aperture marked *w*; and so the color of any electron which is fed in through that aperture on the left can later be inferred from its final position. It's possible to build "hardness boxes" too, and they work in just the same way (see figure 1.2).

Measurements with hardness and color boxes are repeatable, which is something we've grown accustomed to requiring, by definition, of a "good" measurement of a bona fide physical variable. If, say, a certain electron is measured with a color box to be black, and if that electron (without having been tampered with in the meantime) is subsequently fed into the left aperture of another

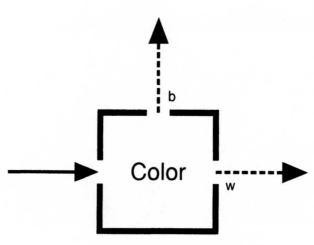

Figure 1.1

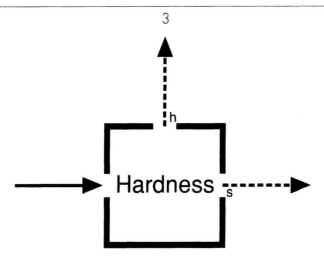

Figure 1.2

color box, then that electron will with certainty emerge from that second color box through the *b* aperture as well. The same goes for white electrons, and the same goes (with hardness boxes) for hard and soft electrons too. All that can be (and has been) confirmed by means of tests with those boxes.

Now, suppose that it occurs to us to be curious about the possibility that the color and hardness properties of electrons might somehow be related to one another. One way to look for such a relation might be to check for *correlations* between the values of the hardness and color properties of electrons. It's easy to check for correlations like that with our boxes; and it turns out (once the checking is done) that no such correlations exist. Of any large collection of, say, white electrons, all of which are fed into the left aperture of a hardness box, precisely half emerge through the hard aperture, and precisely half emerge through the soft one. The same goes for black electrons fed into the left aperture of a hardness box, and the same for hard or soft ones fed into the left apertures of color boxes. The color (hardness) of an electron apparently entails nothing whatever about its hardness (color).

Suppose we set up a sequence of three boxes. First a color box, say, then a hardness box, and then another color box. Consider an electron which emerges through the white aperture of the first color box, and thereafter (without having been tampered with)[3] is fed into the left aperture of the hardness box, and which happens to emerge from *that* box through the soft aperture (as half of such electrons will), and thereafter (once again with no tampering) is fed into the left aperture of the second color box. That electron, as it enters that third box, is presumably known to be both white and soft. As there has been no tampering between boxes here, we should expect that the electron will emerge from this third box through the white aperture, confirming the result of the first measurement. As a matter of fact, that isn't what happens. Precisely half of such electrons emerge from the white aperture of that third box, and the other half (the other half, that is, of those electrons which have been measured to be white and soft by the previous two boxes) emerge from the black aperture. The same goes for any other pair of results in the first two boxes, and the same goes if the color boxes in the above example are replaced with hardness boxes and the hardness box with a color one. Apparently (in the example we considered) the presence of the hardness box between the two color boxes *itself* constitutes some sort of color tampering. Indeed, that hardness box must be what's to blame for changing half of those white electrons to black ones, since we already know that two color measurements, without tampering between the boxes and without an intervening hardness measurement, will invariably produce identical results!

Perhaps the hardness box is poorly built, *crudely* built. It seems to do its job of measuring hardness (without disturbing the hardness in that process) well; but in the course of doing that job it apparently does disrupt color. That raises two questions. First, whether hardness boxes can be built less crudely; whether the job of measuring hardness can be accomplished more delicately, whether it can be accomplished without disrupting color. Second, in the case of this "crude" apparatus, this apparatus which changes the colors of fully half of the electrons whose hardnesses it mea-

---

3. Exactly what constitutes "tampering" and what doesn't is (of course) something one learns, at first, by *experience*.

sures: what is it that determines precisely *which* electrons have their colors changed and which don't?

Let's talk about the second question first. The right way to discover precisely what it is that determines which electrons change color in passing through that intermediate hardness box and which don't would seem to be to monitor very carefully all of the measurable properties of all of the electrons which are fed into that first color box in the course of some particular experiment and which are at that point found to be, say, white; and to make very certain that the physical states of those three boxes are held perfectly and constantly fixed throughout that experiment; and to look for correlations between the measurable physical properties of those incoming electrons and their final positions as they emerge from the second color box. Well, it turns out that, in so far as we are presently able to tell, absolutely no such correlations exist. As a matter of fact, when we take whatever pains we know how to take to insure that all of the electrons in some particular experiment are fed into that first color box with precisely identical sets of physical properties, and to insure that the physical states of those boxes are indeed held precisely and constantly fixed throughout that experiment, the statistics of final outcomes remain precisely as they were described above. In so far as we are now able to determine, then, this second question has no answer. That is, in so far as we are now able to determine, those electrons whose color is changed by passage through a hardness box and those electrons whose color isn't changed by passage through a hardness box need not initially differ from one another in any way whatever.

Let's try the first question. Can hardness boxes be built less crudely? Well, hardness boxes can be built in a number of entirely different ways. We can try each one. It turns out that they all produce the statistics described above. All of those boxes change the color of (statistically) precisely half of the electrons which pass through them. We can try to be much more careful and much more precise in constructing our hardness boxes, but it turns out that that doesn't change anything either. What's striking here isn't that we are unable to build hardness boxes which don't disturb the color of electrons at all, but rather that we are unable to move the statistics of color disruption even so much as one millionth of one percentage point away from fifty-fifty, in either direction, no matter what we try. So long as the device at hand fulfills the *definitional* requirements of a hardness box (that is: so long as it's a device with which the hardnesses of electrons can be determined, repeatably), then the color randomization produced by that device has always, in our experience, been total; and all of the same goes for the effects of color boxes on hardness.

Suppose we wanted to build a color-*and*-hardness box; that is, a device with which both the color *and* the hardness of electrons could be determined. That box would need five apertures (see figure 1.3); one (on the left) where the electrons are taken in, one where white and hard electrons emerge, one where white and soft ones emerge, one where black and hard ones emerge, and one where black and soft ones emerge.

Consider how we could build a box like that. A box like that would seem to need to consist of a hardness box and a color box. But if the incoming electrons are made to pass first through, say, the hardness box, then their hardnesses might subsequently be

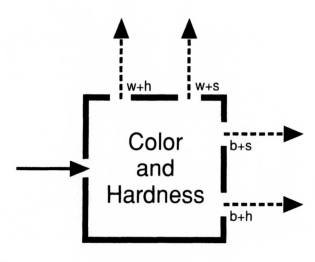

Figure 1.3

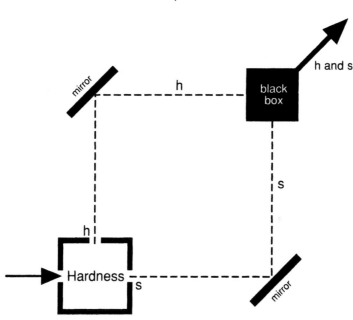

Figure 1.4

changed when they pass through the color box, and we would end up with reliable information only about the colors of the emergent electrons. If we put the color box first, we would end up with reliable information only about the hardnesses. Nobody's been able to think of any *other* ways to build a color-and-hardness box, and it's hard to imagine how, in principle, there could be other ways (other, that is, than building them out of color boxes and hardness boxes). So the task of putting ourselves in a position to say "the color of this electron is now such-and-such and the hardness of this electron is now such-and-such" seems to be fundamentally beyond our means.

That fact is an example of the *uncertainty principle*. Measurable physical properties like color and hardness are said to be "incompatible" with one another, since measurements of one will (so far as we know) always necessarily disrupt the other.

O.K. Let's get in deeper. Consider the rather complicated device shown in figure 1.4. In one corner there's a hardness box. Hard electrons emerge from that box, along route *h*, and at a certain point on route *h* there's a "mirror," or a "reflecting wall," which changes the direction of motion of the electron but doesn't change anything else (more particularly, it doesn't change the hardness of an electron that bounces off it), as shown. Similarly, soft electrons emerge along route *s*, and they run into the same sort of mirror, and finally routes *h* and *s* converge at the "black box."

That black box is another device for changing the directions of motion of electrons on either of these two routes without changing their hardnesses. What it does is to make those two routes coincide after they pass through it. The fact that a soft electron entering the black box along *s* will emerge along *h and s* as a soft electron, and that a hard electron entering the black box along *h* will emerge along *h and s* as a hard electron, can, of course, be tested independently before we set everything up; and the same goes for the mirrors.

Let's do some experiments with this device.

Suppose, first, that we feed a white electron into the hardness box, and then, once it has passed all through the setup drawn in figure 1.4, along path *h and s*, we measure its hardness. What will we expect to find? Well, 50 percent of such electrons will take route *h* out of the hardness box, and 50 percent will take route *s*. And so 50 percent will end up at *h and s* as hard electrons, and 50 percent will end up there as soft ones (since nothing between the hardness box and *h and s*, on either route, can change hardnesses). On repeating the experiment many times, we find that that is precisely what happens.

Suppose that we feed a hard electron into the hardness box and then measure its color at *h and s*. Well, every electron like that will take route *h* through the apparatus and will emerge at *h and s* as a hard electron, and, as we know, 50 percent of such electrons will be found to be black by a color measurement, and 50 percent will be found to be white. If a soft electron were fed into the hardness box at the beginning, it would take route *s* and still be soft at *h and s*, and so the final color-measurement statistics of electrons like *that* ought to be fifty-fifty too. And all that, as before, is indeed just what happens when we actually carry out a large number of such experiments.

Let's try one more. This one is the surprise. Suppose that we feed a white electron into the hardness box at the beginning, and measure color at the end. What should we expect then? Well, just as above, 50 percent of such electrons will turn out to be hard and will take *h* through the apparatus; and 50 percent will turn out to be soft and will take *s*. Consider the first half. They will all emerge at *h and s* as hard electrons, and consequently 50 percent will turn out to be black, on a color measurement, and 50 percent will turn out to be white. The second half, on the other hand, will all emerge as soft electrons, but of course their color statistics will be precisely the same. Putting all that together, it follows that of any large set of white electrons fed (one at a time) into this apparatus, half should be found at the end to be white, and the other half should be found to be black; and of course that makes very good sense, since (aside from a few harmless mirrors and a harmless black box) this apparatus is really just a hardness box, which is already known to randomize the color!

The funny thing is that when you *try* this last experiment, that isn't what happens at all! It turns out that exactly 100 percent of white electrons fed into this apparatus come out white at the end.

That's very odd. It's hard to imagine what can possibly be going on. Maybe another experiment will help clear things up. Let's try this: Rig up a small, movable, electron-stopping wall that can be slid, at will, in and out of, say, route *s* (see figure 1.5). When the wall is "out," we have precisely our earlier apparatus; but when the wall is "in," all electrons moving along *s* get stopped and only those moving along *h* get through to *h and s*.

What should we expect to happen when we slide the wall in? Well, to begin with, the overall output of electrons at *h and s* ought to go down by 50 percent, since the input white electrons ought to be half hard and half soft, and the latter shouldn't now be getting through. What about the color statistics of that remaining 50 percent? Well, when the wall is out, 100 percent of white electrons fed in end up white. That means that all the electrons that take *s* end up white and all the electrons that take *h* end up white; and

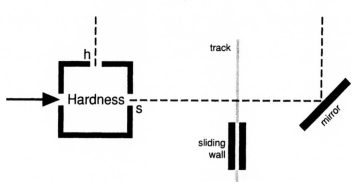

Figure 1.5

since we can easily verify that whether the wall is in or out of *s* can have no effect on the colors of electrons traveling along *h*, that implies that those remaining 50 percent should all be white.

What actually happens when we do the experiment? Well, the output is down by 50 percent, as we expect. But the remaining 50 percent is not all white. It's half white and half black. The same thing happens, similarly contrary to our expectations, if we insert a wall in the hard path instead.

Now we're in real trouble.

Consider an electron which passes through our apparatus when the wall is out. Consider the possibilities as to which route that electron can have taken. Can it have taken *h*? Apparently not, because electrons which take *h* (as we've just seen again) are known to have the property that their color statistics are fifty-fifty, whereas an electron passing through our device with the wall out is known to have the property of being white at *h and s*! Can it have taken *s*, then? No, for the same reasons. Can it somehow have taken *both* routes? Well, suppose that when a certain electron is in the midst of passing through this apparatus, we stop the experiment and look to see where it is. It turns out that half the time we find it on *h*, and half the time we find it on *s*. We never find two electrons in there, or two halves of a single, split electron, one on each route, or anything like that. There isn't any sense in which the electron seems to be taking both routes. Can it have taken neither route?

Certainly not. If we wall up *both* routes, nothing gets through at all.

So what we're faced with is this: Electrons passing through this apparatus, in so far as we are able to fathom the matter, do not take route *h* and do not take route *s* and do not take both of those routes and do not take neither of those routes; and the trouble is that those four possibilities are simply all of the logical possibilities that we have any notion whatever of how to entertain!

What can such electrons be doing? It seems they must be doing something which has simply never been dreamt of before (if our experiments are valid, and if our arguments are right). Electrons seem to have modes of being, or modes of moving, available to them which are quite unlike what we know how to think about.

The name of that new mode (which is just a name for something we don't understand) is *superposition*. What we say about an initially white electron which is now passing through our apparatus with the wall out is that it's not on *h* and not on *s* and not on both and not on neither, but, rather, that it's in a superposition of being on *h and* being on *s*. And what that means (other than "none of the above") we don't know. And some of what this book is going to be about are a number of attempts to (as it were) say something more about superposition than that.

Let's make the main point (which is that superpositions are *extraordinarily* mysterious situations) one or two more times.

Here's a second example. It's possible to construct boxes which I'd like to call "total-of-nothing" boxes. A total-of-nothing box is a box with two apertures. An electron which is fed into one aperture emerges from the other with all of its measurable properties (color, hardness, velocity, whatever) *unchanged;* and the time of passage through the box for any electron is equal to the time it would have taken for that electron to traverse an empty space the size of the box. Those are the defining properties of total-of-nothing boxes.

Clearly, there are lots of ways to build total-of-nothing boxes. A completely empty box with two holes in it is a total-of-nothing box. We can also imagine building boxes which do all sorts of violent things to the electrons fed into them but which subsequently *undo* all those things and finally eject those electrons, at the right times, at the right speeds, with all of their original physical properties. Every box like that will be a total-of-nothing box too.

Now, recall the two-path apparatus of figure 1.4. White electrons fed into that apparatus always come out white. It turns out to be possible to build a total-of-nothing box which, when inserted into either one of those two paths, will make all of those outgoing electrons black instead of white. If the box is removed from the path, the outgoing electrons will all go back to being white.[4] So, inserting such a box into one of those paths will change the color of an electron passing through this two-paths apparatus. But total-of-nothing boxes, by definition, change none of the properties of electrons which pass through them; and, of course, total-of-nothing boxes change none of the properties of electrons which *don't* pass through them! So it isn't possible that these electrons pass through the total-of-nothing box on one of the paths, since in that event their colors couldn't have been changed from white to black by the presence of the box; and it isn't possible that those electrons pass outside of the box, since in *that* event their colors couldn't have been changed either! And it isn't possible (by our earlier arguments) that those electrons pass both inside and outside of the box, and it isn't possible that they pass neither inside *nor* outside of the box.

Here's one final example, a very well-known one. Consider an arrangement such as is depicted in figure 1.6. On the left is a source of electrons. Electrons emerge from that source in a whole spectrum of possible directions, as shown. Slightly farther to the right is a screen which electrons cannot pass through, and that screen has two holes in it. Still farther to the right is a fluorescent screen, much like a television screen, which lights up, at the point of impact, whenever it is struck by an electron (that is, this fluorescent screen is a *measuring device* for the positions of electrons).

Suppose, first, that the top hole in the first screen is closed up, as in figure 1.6A. Electrons emerge, one by one, from the source,

4. That there can be boxes like that was first predicted in a famous paper of Aharonov and Bohm (1959).

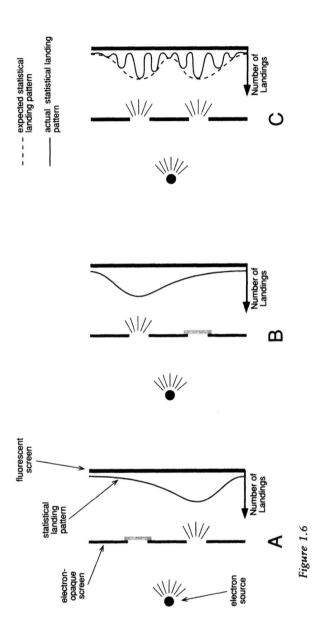

Figure 1.6

and move toward the first screen. Most of them run into the screen and are stopped there. Some get through the hole. Those latter land at various points on the fluorescent screen. The statistics of those landings (that is: how many land in any particular region) are shown in the figure. Figure 1.6B gives the same information for the case when the bottom hole is closed instead of the top one.

What sort of landing statistics should we expect when *both* holes are open? Well, all of the electrons which arrive at the fluorescent screen will have passed either through the top hole or through the bottom one. Those which pass through the bottom hole are known (by our first experiment with this apparatus) to give rise to the statistical landing pattern of figure 1.6A; and those which pass through the top hole are known to give rise to the statistical landing pattern of figure 1.6B; and so, in the event that both holes are open (and in the event that only one electron is allowed to pass through this apparatus at a time), we should expect a statistical landing pattern which is the direct sum of those two (as shown in figure 1.6C). But that (it will be no surprise by now) is not what happens. The statistical landing pattern which emerges on the fluorescent screen when both holes are open is markedly different from the direct sum of patterns A and B. So, it's inconsistent with these experimental results to suppose that an electron passing through this apparatus passes through the upper hole when both holes are open; and it is inconsistent with these experimental results to suppose that an electron passing through this apparatus passes through the lower hole when both holes are open. And the same sorts of experiments and arguments as were described above will entail that it also can't be maintained that such electrons pass through both holes, and that it also can't be maintained that they pass through neither hole.

These electrons are (then) in superpositions of passing through the upper hole and passing through the lower one; but (once again) we have no idea, or rather only a negative idea, of what that means.

All this stuff about superpositions, by the way, sheds a very curious light on the phenomena of uncertainty and incompatibility (between color and hardness, say) that we ran into at the beginning of this chapter.

Consider this: We know, by experiment, that electrons emerge from the hard aperture of a hardness box if and only if they're hard electrons when they enter that box (that, as a matter of fact, is the sole property in virtue of which anything ever deserves to be *called* a hardness box). Similarly for soft ones. Now, when a *white* electron is fed into a hardness box, it invariably emerges (as we've just seen) neither through the hard aperture nor through the soft one nor through both nor through neither. So, it follows that a white electron can't *be* a hard one, or a soft one, or (somehow) both, or neither. To say that an electron is white must be just the same as to say that it's in a *superposition* of being hard and soft.

And consider (in light of that) why it is that we find that we can't ever put ourselves in a position to say, "The color of this electron is now such-and-such and its hardness is now such-and-such." It isn't that our color and hardness boxes are built (somehow) crudely (which is what we suspected at first). And in fact it isn't *at all* a matter of our being unable to simultaneously *know* what the color and the hardness of a certain electron is (that is: it isn't a matter of *ignorance*). It's deeper than that. It's that any electron's even *having* any definite color apparently entails that it's neither hard nor soft nor both nor neither, and that any electron's even having any definite hardness apparently entails that it's neither black nor white nor both nor neither.

And consider (while we're at it) why it is that the rules for predicting the outcome of a measurement of (say) the hardness of a white electron turn out (in so far as we're now able to determine) to be probabilistic rules rather than deterministic ones. It's like this: if it could ever be said of a white electron that a measurement of its hardness will with certainty produce the outcome (say) "soft," or if it could ever be said of a white electron that a measurement of its hardness will with certainty come out "hard," that would apparently be inconsistent with what we now *know* to be the case, which is that such an electron (a white one) is in a *superposition* of being hard and being soft. And on the other hand, our experience is that every hardness measurement whatsoever either comes out "hard" or it comes out "soft." And so apparently the outcome of a hardness measurement on a white electron has *got* to be a matter of *probability*.

But of course the business of talking more carefully about all this stuff (which is what I want to do here) will require an appropriate mathematical apparatus. And so the next chapter will lay an apparatus like that out.

The Mathematical Formalism and the Standard Way of Thinking about It

There is an algorithm (and the name of that algorithm, of course, is quantum mechanics) for predicting the behaviors of physical systems, which correctly predicts all of the unfathomable-looking behaviors of the electron in the story in Chapter 1, and there is a standard way of interpreting that algorithm (that is, a way attempting to fathom those behaviors, a way of attempting to confront the fact of superposition) which can more or less be traced back to some sayings of Niels Bohr.[1] This chapter will describe that algorithm and rehearse that standard way of talking about it, and then it will apply them both, in some detail, to that story.

Mathematical Preliminaries

Let me say a few things, to begin with, about the particular mathematical language in which it is most convenient to write the algorithm down.

Let's start with something about vectors. A good way to think about vectors is to think about arrows. A vector is a mathematical object, an abstract object, which (like an arrow) is characterized by

---

1. The story of the evolution of this standard way of thinking is a very long and complicated one, and it will be completely ignored here. The far more obscure question of what Bohr himself really thought about these issues will be ignored too. What will matter for us is the legacy which Bohr and his followers have left, by whatever route, and whatever they themselves may have originally thought, to modern physics. That legacy, as it stands now, can be characterized fairly clearly. The name of that legacy is the Copenhagen interpretation of quantum mechanics.

a direction (the direction in which the arrow is pointing) and a magnitude (the length of the arrow).

Think of a coordinate system with a specified origin point. Every distinct geometrical point in the space mapped out by such a coordinate system can be associated with some particular (and distinct) vector, as follows: the vector associated with any given point (in that given coordinate system) is the one whose tip lies at the given point and whose tail lies at the origin. The length of that vector is the distance between those two points, and the direction of that vector is the direction from the origin to the given point (see figure 2.1).

The infinite collection of vectors associated with *all* the points in such a space is referred to as a *vector space*.

Spaces of points can be characterized by (among other things) their dimensionality, and spaces of vectors can too. The dimension of a given vector space is just the dimension of the associated space of points. That latter dimension, of course, is equal to the number of magnitudes, the number of coordinates, that need to be specified in order to pick out (given a coordinate system) some particular geometrical point.

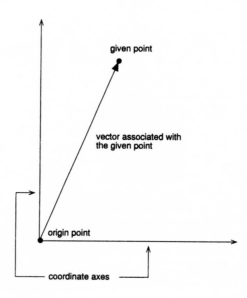

Figure 2.1

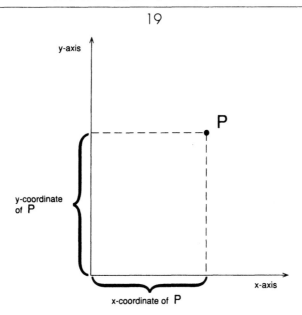

Figure 2.2

Figure 2.2, for example, shows a two-dimensional space, a *plane* of points, wherein (given the indicated coordinate system) two coordinates need to be specified (the *x*-coordinate and the *y*-coordinate) in order to pick out a point. The reader can convince herself that a line of points forms a one-dimensional space, and that the space we move around in has three dimensions. Spaces of points with more dimensions than that are hard to visualize, but the formal handling (that is: the mathematical handling) of such spaces is not a problem.

Let's introduce a notation for vectors: the symbols | ⟩ around some expression will henceforth indicate that that expression is the name of the vector; so that, for example, |A⟩ will denote the vector called A. That's the notation most commonly used in the literature of quantum mechanics.

Vectors can be added to one another. Here's how: To add |A⟩ to |B⟩, move the tail of |B⟩ to the tip of |A⟩ (without altering the length or the direction of either in the process). The sum of |A⟩ and |B⟩ (which is written |A⟩ + |B⟩) is defined to be that vector (|C⟩) whose tail now coincides with the tail of |A⟩ and whose tip now coincides with the tip of $|B\rangle$ (see figure 2.3). The sum of any two vectors in any particular vector space is always another vector in that same space (that, indeed, is part of the definition of a vector space). Think, for example, of the spaces discussed above.

That fact is going to be important. Vectors, in quantum mechanics, are going to represent physical states of affairs. The *addition* of vectors will turn out to have something to do with the *superposition* of physical states of affairs. The fact that two vectors can be added together to form a third will turn out to accommodate, within the algorithm, the fact that certain physical states of affairs, states like being white, are superpositions of certain other states of affairs, states like being hard and being soft; but of all this more later.

Vectors can be multiplied too. There are two ways to multiply them. First of all, they can be multiplied by numbers. The vector $5|A\rangle$, say, is defined to be that vector whose direction is the same as the direction of $|A\rangle$ and whose length is 5 times the length of $|A\rangle$. $5|A\rangle = |A\rangle + |A\rangle + |A\rangle + |A\rangle + |A\rangle$. Of course, if $|A\rangle$ is an element of a certain vector space, any number *times* $|A\rangle$ will be an element of that space too.

The other way to multiply vectors is to multiply them by other vectors. The multiplication of a vector by another vector yields a number (not a vector!). $|A\rangle$ times $|B\rangle$ (which is written $\langle A|B\rangle$) is defined to be the following number: the length of $|A\rangle$ times the length of $|B\rangle$ times the cosine of the angle, θ, between $|A\rangle$ and $|B\rangle$.

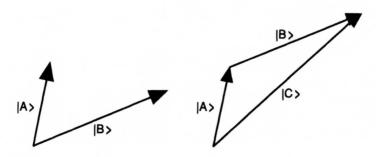

Figure 2.3

The length of $|A\rangle$ (also called the *norm* of $|A\rangle$, which is written $|A|$) is obviously equal to the square root of the number $\langle A|A\rangle$, since the cosine of $0°$ ($0°$ is the angle between $|A\rangle$ and itself) is equal to 1.

So, vectors plus vectors are vectors, and vectors times numbers are vectors, and vectors times vectors are numbers.

Here's a slightly more sophisticated way of defining a vector space: a vector space is a collection of vectors such that the sum of any two vectors in the collection is also a vector in the collection, and such that any vector in the collection times any (real) number is also a vector in the collection. Such collections (by the way) clearly have to be infinite. Think, again, of the examples of spaces described above.

If $|A| \neq 0$ and $|B| \neq 0$ and yet $\langle A|B\rangle = 0$ (that is: if the angle between $|A\rangle$ and $|B\rangle$ is $90°$, since $\cos 90° = 0$), then $|A\rangle$ and $|B\rangle$ are said to be orthogonal to one another. *Orthogonal* just means perpendicular.

Here's another definition of dimension: The dimension of a vector space is equal (by definition) to the maximum number (call that number N) of vectors $|A_1\rangle$, $|A_2\rangle$, . . . $|A_N\rangle$ which can be chosen in the space such that for all values of i and j from 1 through N such that $i \neq j$, $\langle A_i|A_j\rangle = 0$. That is, the dimension of a space is equal to the number of mutually perpendicular directions in which vectors within that space can point.

Given a space, there are generally lots of ways to pick out those directions. Pick a vector, at random, from an N-dimensional space. It will always be possible to find a set of $N - 1$ other vectors in that space which are all orthogonal to that original vector and to one another. In most cases, given that original vector, there will still be many such orthogonal sets (or, rather, an infinity of such sets) to choose from. Figure 2.4 shows some examples.

Think of an N-dimensional space. Think of any collection of N mutually orthogonal vectors in that space, and suppose that the norm, the length, of each of those vectors happens to be 1. Such a set of vectors is said to form an orthonormal *basis* of that N-dimensional space. *Ortho* is for orthogonal, *normal* is for norm-1, and here's why sets of vectors like that are called *bases* of their spaces: Suppose that the set $|A_1\rangle$, $|A_2\rangle$, . . . $|A_N\rangle$ forms a basis of a

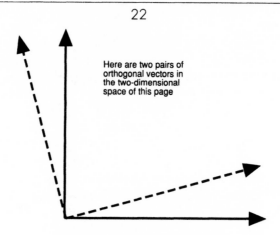

Here are two pairs of
orthogonal vectors in
the two-dimensional
space of this page

All four of these vectors, moreover, are orthogonal
to any vector pointing directly out of the page

Figure 2.4

certain N-dimensional vector space; it turns out that any vector
whatever in that space (call it $|B\rangle$) can be expressed as the following
sort of sum:

$$(2.1) \qquad |B\rangle = b_1|A_1\rangle + b_2|A_2\rangle + \ldots b_N|A_N\rangle$$

where the b_i are all simply numbers—more particularly, simply the
following numbers:

$$(2.2) \qquad b_i = \langle B|A_i\rangle$$

So any vector in a vector space can be "built up" (as in (2.1)) out
of the elements of any basis of that space. All that is illustrated, for
a two-dimensional space, in figure 2.5.

Bases end up amounting to precisely the same thing as coordinate
systems: given a coordinate system for an N-dimensional point
space, N numbers (the coordinate values) will suffice to pick out a
point; given a basis of an N-dimensional vector space, N numbers
(the b_i of equation (2.1)) will suffice to pick out a vector. Vectors
which are of norm 1 and which point along the perpendicular
coordinate axes of an N-dimensional point space will constitute an

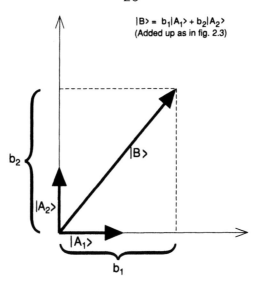

Figure 2.5

orthonormal basis of the associated N-dimensional vector space, and vice versa.

For any space of more than a single dimension, there will be an infinity of equivalently good orthornormal bases to choose from. Any vector in that space will be writable, a la (2.1), in terms of any of those bases, but of course, for a given vector $|B\rangle$, the numbers b_i in (2.1) (which, by the way are called expansion coefficients) will differ from basis to basis. Figure 2.6 shows how that works.

Now, it happens to be the case that for any three vectors $|A\rangle$, $|B\rangle$, and $|C\rangle$, the product $|A\rangle$ times the vector $(|B\rangle + |C\rangle)$ is equal to the product $|A\rangle$ times $|B\rangle$ plus the product $|A\rangle$ times $|C\rangle$:

(2.3) $\langle A\|B\rangle + |C\rangle\rangle = \langle A|B\rangle + \langle A|C\rangle,$

and that can be shown to entail, for any two vectors $|M\rangle$ and $|Q\rangle$, that

(2.4a) $|M\rangle + |Q\rangle = (m_1 + q_1)|A_1\rangle + (m_2 + q_2)|A_2\rangle + \ldots +$
$(m_N + q_N)|A_N\rangle$

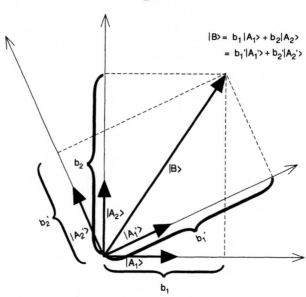

$$|B\rangle = b_1|A_1\rangle + b_2|A_2\rangle$$
$$= b_1'|A_1'\rangle + b_2'|A_2'\rangle$$

Figure 2.6

and that

(2.4b) $\langle M|Q\rangle = m_1q_1 + m_2q_2 + \ldots + m_Nq_N$

wherein the m_i and q_i are the expansion coefficients of $|M\rangle$ and $|Q\rangle$, respectively, in any particular basis $|A_i\rangle$. The numbers q_i and m_i will, of course, depend on the choice of basis, but note that the sum of their products in (2.4b) (which is equal to $\langle M|Q\rangle$, which depends only on which vectors $|M\rangle$ and $|Q\rangle$ happen to *be*, and not on which basis we happen to map them out in) will not. That sum, rather, will be *invariant* under changes of basis.

Suppose that we have agreed to settle on some particular basis for some particular vector space. Once that's done, all that will be required for us to pick out some particular vector ($|Q\rangle$, say) will be to specify the numbers (the expansion coefficients) q_i of $|Q\rangle$ for that particular basis. Those N numbers then (once the basis is chosen) can serve to represent the vector. Those numbers are usually written down in a column; for example:

(2.5)
$$|Q\rangle = \begin{bmatrix} 1 \\ 5 \\ -\frac{3}{2} \end{bmatrix}$$

= the three–dimensional vector for which $\begin{cases} \langle Q|A_1\rangle = 1 \\ \langle Q|A_2\rangle = 5 \\ \langle Q|A_3\rangle = -\frac{3}{2} \end{cases}$

(see equation (2.2)), where the $|A_i\rangle$ are the chosen basis vectors. It follows from (2.4b) that the norm (the length) of any vector $|Q\rangle$ will be equal to the square root of the sum of the squares of its expansion coefficients. That number, too, must obviously be invariant under changes of basis.

That's all that will concern us about vectors. The other sorts of mathematical objects which we shall need to know something about are *operators*.

Operators are mechanisms for making new vectors out of old ones. An operator on a vector space, more particularly, is some definite prescription for taking every vector in that space into some other vector; it is a mapping (for those readers who know the mathematical meaning of that word) of a vector space into itself.

Let's introduce a notation. Suppose that the operator called O is applied to the vector $|B\rangle$ (that is: suppose that the prescription called O is carried out on the vector $|B\rangle$). The result of that operation, of that procedure, is written:

(2.6) $O|B\rangle$

Then what was just said about operators can be expressed like this:

(2.7) $O|B\rangle = |B'\rangle$ for any vector $|B\rangle$ in the vector space on which O is an operator.

where $|B'\rangle$ is some vector in the same space as $|B\rangle$.

Here are some examples. One example is the "unit" operator

(that's the prescription which instructs us to multiply every vector in the space by the number 1, to transform every vector into itself). The unit operator is the one for which

$$(2.8) \qquad O_U|B\rangle = |B\rangle = |B'\rangle$$

Another example is the operator "multiply every vector by the number 5." Another example is the operator "rotate every vector clockwise by 90° about some particular vector $|C\rangle$" (see figure 2.7). Another example is the operator "map every vector in the space into some particular vector $|A\rangle$."

The particular sorts of operators which will play a vital role in the quantum-mechanical algorithm are *linear operators*. Linear operators are, by definition, operators which have the following properties:

$$(2.9a) \qquad O(|A\rangle + |B\rangle) = O|A\rangle + O|B\rangle$$

and

$$(2.9b) \qquad O(c|A\rangle) = c(O|A\rangle)$$

Suppose that $|C\rangle$ is a vector pointing directly out of the page. Then the operator "rotate every vector in the space clockwise by 90° about $|C\rangle$" will do this to $|A\rangle$ and $|B\rangle$:

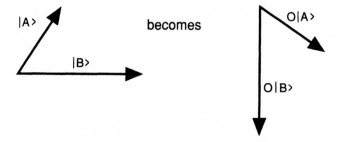

becomes

Figure 2.7

for any vectors $|A\rangle$ and $|B\rangle$ and any number c. Here's what (2.9a) says: take that vector which is the sum of two other vectors $|A\rangle$ and $|B\rangle$ (such sums, remember, are always vectors), and operate on that sum with any linear operator. The resultant (new) vector will be that vector which is the sum of the new vector produced by operating on $|A\rangle$ with O and the new vector produced by operating on $|B\rangle$ with O. What (2.9b) says is that the vector produced by operating on c times $|A\rangle$ with O is the same as c times the vector produced by operating on $|A\rangle$ itself with O, for any number c.[2]

Now, the two conditions in (2.9) pick out a very particular sort of operator. They are by no means properties of operators in general. Let me leave it as an exercise for the reader to show, for example, that of the four operators just now described, the first three are linear and the last one isn't.

Linear operators are very conveniently representable by arrays of numbers. We learned it was possible, remember, to represent any N-dimensional *vector*, given a choice of basis, by N *numbers* (a la equation (2.5)); and it similarly turns out to be possible to represent any linear operator (the linearity is crucial here) on an N-dimensional vector space by N^2 numbers. Those N^2 numbers are traditionally arranged not in a column (as in equation (2.5), for vectors), but in a *matrix*, as (for a two-dimensional operator, say) follows:

$$(2.10) \qquad O = \begin{bmatrix} O_{11} & O_{12} \\ O_{21} & O_{22} \end{bmatrix}$$

The numbers O_{ij} in (2.10) are defined to be

$$(2.11) \qquad O_{ij} = \langle A_i | O | A_j \rangle\rangle$$

That is: the number O_{ij} is the vector $O|A_j\rangle$ multiplied by the vector $|A_i\rangle$ (such products of vectors, remember, are always *numbers*),

---

2. The two parts of (2.9) aren't completely independent of one another, by the way. Note, for example, that in the event that c is an integer, (2.9b) is entailed by (2.9a).

where the $|A_N\rangle$ are the chosen basis vectors of the space. There's a rule for multiplying operator matrices by vector columns, which is:

$$(2.12) \quad \begin{bmatrix} O_{11} & O_{12} \\ O_{21} & O_{22} \end{bmatrix} \times \begin{bmatrix} b_1 \\ b_2 \end{bmatrix} = \begin{bmatrix} (O_{11}b_1 + O_{12}b_2) \\ (O_{21}b_1 + O_{22}b_2) \end{bmatrix}$$

Note that the right-hand side of (2.12) is a vector column; so this rule stipulates that the product of an operator matrix and a vector column is a new vector column.

Here's why all this notation is useful: it turns out (we won't prove it here) that any linear operator whatever can be uniquely specified (given a basis choice) by specifying the N^2 O_{ij} of equations (2.10) and (2.11) (just as any *vector* can be uniquely specified by specifying the N b_i of equations (2.1) and (2.2) and (2.5)); and it turns out that for any linear operator O, we can calculate O's effect on any vector $|B\rangle$ simply by multiplying the O-matrix by the $|B\rangle$-column (given, as always, a basis choice) as in (2.12). That is, for any linear operator O and any vector $|B\rangle$:

$$(2.13) \quad |B'\rangle = \begin{bmatrix} O_{11} & O_{12} \\ O_{21} & O_{22} \end{bmatrix} \times \begin{bmatrix} b_1 \\ b_2 \end{bmatrix} = \begin{bmatrix} (O_{11}b_1 + O_{12}b_2) \\ (O_{21}b_1 + O_{22}b_2) \end{bmatrix}$$

$$= \underbrace{(O_{11}b_1 + O_{12}b_2)}|A_1\rangle + \underbrace{(O_{21}b_1 + O_{22}b_2)}|A_2\rangle = |B'\rangle$$

these are *numbers*

where $|A_i\rangle$ are the chosen basis vectors.[3] (The next-to-last equality follows from equations (2.1) and (2.2) and (2.5)).

3. Perhaps it's worth saying all that out in words: In order to calculate the effect of any linear operator O on any vector $|B\rangle$, first choose a basis, then calculate the $|B\rangle$ column vector in that basis by means of formula (2.2); then calculate the operator matrix in that basis by means of formula (2.11); then multiply that column vector by that operator matrix by means of formula (2.12); and the result of that multiplication will be the column vector, in that same basis, of the new vector $|B'\rangle$ (that is, the vector obtained by operating with O on $|B\rangle$).

THE MATHEMATICAL FORMALISM

One more definition will be useful. If it happens to be the case for some particular operator O and some particular vector $|B\rangle$ that

(2.14) $O|B\rangle = @|B\rangle$

where @ is some number—that is, if the new vector generated by operating on $|B\rangle$ with O happens to be a vector pointing in the same direction as $|B\rangle$—then $|B\rangle$ is said to be an *eigenvector* of O, with *eigenvalue* @ (where @ is the length of that new vector relative to the length of $|B\rangle$).

Certain vectors will in general be eigenvectors of some operators and not of some others; certain operators will in general have some vectors, and not others, as eigenvectors, and other operators will have *other* vectors as eigenvectors. The operator-eigenvector relation, however, depends only on the vector and the operator in question, and not at all on the basis in which we choose to write those objects down. In other words, if the eigenvector-operator relation obtains between the vector column and the operator matrix of a certain vector and a certain operator in a certain particular basis, then it can be shown that the same relation, with the same eigenvalue, will obtain between the vector column and the operator matrix in any basis whatever of that space.

Here are some examples: all vectors are eigenvectors of the unit operator, and all have eigenvalue 1; and similarly (but with eigenvalue 5) for the operator "multiply every vector by 5." All vectors of the form $@|C\rangle$, where @ is any number, are eigenvectors of the operator "rotate every vector about $|C\rangle$ by 90°"; all those vectors have eigenvalue 1, and there are no other eigenvectors of that operator. The four-dimensional space operator (written down in some particular basis)

(2.15)
$$O = \begin{bmatrix} 5 & 0 & 0 & 0 \\ 0 & \frac{3}{2} & 0 & 0 \\ 0 & 0 & 2 & 0 \\ 0 & 0 & 0 & -7 \end{bmatrix}$$

has eigenvectors (written in that same basis)

$$(2.16) \qquad |A\rangle = \begin{bmatrix} 1 \\ 0 \\ 0 \\ 0 \end{bmatrix} \quad |B\rangle = \begin{bmatrix} 0 \\ 1 \\ 0 \\ 0 \end{bmatrix} \quad |C\rangle = \begin{bmatrix} 0 \\ 0 \\ 1 \\ 0 \end{bmatrix} \quad |D\rangle = \begin{bmatrix} 0 \\ 0 \\ 0 \\ 1 \end{bmatrix}$$

with eigenvalues 5, $3/2$, 2, and -7, respectively. Any number *times* $|A\rangle$ or $|B\rangle$ or $|C\rangle$ or $|D\rangle$ will be an eigenvector of O too, with the same eigenvalue; but vectors like $|A\rangle + |B\rangle$ won't be eigenvectors of O.

Quantum Mechanics

Now we're in a position to write out the algorithm. It pretty much all boils down to five principles.

(A) Physical States. Physical situations, physical states of affairs, are represented in this algorithm by vectors. They're called *state vectors.* Here's how that works: Every physical system (that is: every physical object, and every collection of such objects), to begin with is associated in the algorithm with some particular vector space; and the various possible physical states of any such system are stipulated by or correspond to vectors, and more particularly to vectors of length 1, in that system's associated space; and every such vector is taken to pick out some particular such state; and the states picked out by all those vectors are taken to comprise all of the possible physical situations of that system (the correspondence isn't precisely one-to-one, however: we shall soon discover, for example, that for any vector $|A\rangle$ of length 1, $-|A\rangle$ must necessarily pick out the same physical state as $|A\rangle$ does).

This will turn out to be a very apt way to represent states, since (as I mentioned before) the possibility of "superposing" two states to form another gets reflected in the algorithm by the possibility of adding (or subtracting) two vectors to form another.

(B) Measurable Properties. Measurable properties of physical systems (such properties are referred to as *observables,* in the quan-

tum-mechanical literature) are represented in the algorithm by linear operators on the vector spaces associated with those systems. There's a rule that connects those operators (and their properties) and those vectors (and their physical states), which runs as follows: If the vector associated with some particular physical state happens to be an eigenvector, with eigenvalue (say) a, of an operator associated with some particular measurable property of the system in question (in such circumstances, the state is said to be an "eigenstate" of the property in question), then that state has the value a of that particular measurable property.

Let's try all that out. Let's construct a vector space in which the state of being hard and the state of being soft can be represented. Suppose we let the following two two-dimensional column vectors stand for hardness and softness:

(2.17)
$$|\text{hard}\rangle = \begin{bmatrix} 1 \\ 0 \end{bmatrix} \qquad |\text{soft}\rangle = \begin{bmatrix} 0 \\ 1 \end{bmatrix}$$

Notice that if we adopt (2.17), $\langle \text{hard}|\text{soft}\rangle = 0$ (see equations (2.4) and (2.5)). As a matter of fact, the two vectors in (2.17) constitute a basis of the two-dimensional space which they inhabit. That particular basis, by the way, is precisely the one in which the vector columns in (2.17) have been written down (that is: the relevant basis vectors $|A_1\rangle$ and $|A_2\rangle$ of equation (2.5) are, in the case of (2.17), precisely $|\text{hard}\rangle$ and $|\text{soft}\rangle$).

What operator should represent the hardness property? Let's try this:

(2.18)
$$\text{hardness operator} = \begin{bmatrix} 1 & 0 \\ 0 & -1 \end{bmatrix}$$

where we stipulate that "hardness = +1" means "hard"
and that "hardness = −1" means "soft"

So far all this works out right: $|\text{hard}\rangle$ and $|\text{soft}\rangle$ of equation (2.17) are, indeed, as the reader can now easily confirm, eigenvectors of the hardness operator of equation (2.18), with the appropriate eigenvalues.

Let's push this example further. Remember that it seemed to us in Chapter 1 that the "black" and "white" states must both be superpositions of both of the "hard" and "soft" states; and remember (from the present chapter) that the superposition of physical states is supposed to correspond somehow to the addition or subtraction of their respective state vectors; and remember that the sum or the difference of any two vectors in any particular vector space is necessarily yet another vector in that same space. All that suggests that the states of being white and being black ought to be representable by vectors in this space too, and that there ought to be a color operator on this space. Let's try this one, written down in the basis of equation (2.17):

(2.19)
$$|black\rangle = \begin{bmatrix} 1/\sqrt{2} \\ 1/\sqrt{2} \end{bmatrix} \qquad |white\rangle = \begin{bmatrix} 1/\sqrt{2} \\ -1/\sqrt{2} \end{bmatrix}$$

$$color\ operator = \begin{bmatrix} 0 & 1 \\ 1 & 0 \end{bmatrix} \qquad \begin{array}{l} \text{"color} = +1\text{" means "black"} \\ \text{"color} = -1\text{" means "white"} \end{array}$$

That works out right too: The reader can show that the various stipulations of (2.19) are all consistent with one another, as are the stipulations of (2.17) and (2.18). Furthermore $\langle black|white\rangle = 0$ too; and $|black\rangle$ and $|white\rangle$ constitute another basis of this space.

Now, it follows from (2.4a) that:

(2.20) $\quad$ if $\quad |A\rangle = \begin{bmatrix} a \\ b \end{bmatrix} \quad$ and $\quad |B\rangle = \begin{bmatrix} c \\ d \end{bmatrix}$

in some particular basis, then

$$|A\rangle + |B\rangle = \begin{bmatrix} (a + c) \\ (b + d) \end{bmatrix}$$

in that same basis (and the same applies, of course, to vector columns of any dimension).

Notice, then, how beautifully (2.17) and (2.18) and (2.19) reflect the principles of superposition and incompatibility. First of all, it follows from (2.17) and (2.19) that:

(2.21) $|\text{black}\rangle = \frac{1}{\sqrt{2}}|\text{hard}\rangle + \frac{1}{\sqrt{2}}|\text{soft}\rangle$

$|\text{white}\rangle = \frac{1}{\sqrt{2}}|\text{hard}\rangle - \frac{1}{\sqrt{2}}|\text{soft}\rangle$

$|\text{hard}\rangle = \frac{1}{\sqrt{2}}|\text{black}\rangle + \frac{1}{\sqrt{2}}|\text{white}\rangle$

$|\text{soft}\rangle = \frac{1}{\sqrt{2}}|\text{black}\rangle - \frac{1}{\sqrt{2}}|\text{white}\rangle$

So sums and differences of vectors, in the algorithm, *do* denote *superpositions* of physical states; and (just as we concluded in the last chapter) states of definite color *are* superpositions of different hardness states, and states of definite hardness *are* superpositions of different color states.

Moreover, look how well the forms of the hardness and color operators confirm all this: It's easy to verify that the "black" and "white" vectors aren't eigenvectors of the hardness operator, and that the "hard" and "soft" vectors aren't eigenvectors of the color operator. The hardness and color operators are (just as they ought to be) incompatible with one another, in the sense that states of definite hardness (that is: states whose vectors are eigenvectors of the hardness operator) apparently have no assignable color value (since those vectors aren't eigenvectors of the color operator) and vice versa.

So it turns out that the descriptions of color and of hardness and of all the relations between them can be subsumed within a single, two-dimensional vector space. That space is referred to within the quantum-mechanical literature as the *spin* space, and color and hardness are referred to as *spin* properties.

Let's get back to the enumeration of the five principles.

(C) Dynamics. Given the state of any physical system at any "initial" time (given, that is, the vector which represents the state of that system at that time), and given the forces and constraints to

which that system is subject, there is a prescription whereby the state of that system at any *later* time (that is, the vector at any later time) can, in principle, be calculated. There is, in other words, a *dynamics* of the state vector; there are deterministic laws about how the state vector of any given system, subject to given forces and constraints, changes with time. Those laws are generally cast in the form of an equation of motion, and the name of that equation, for nonrelativistic systems, is the *Schrödinger* equation.

Since every state vector must, by definition, be a vector of length one, the changes in state vectors dictated by the dynamical laws are exclusively changes of direction, and never of length.

Here's an important property of the quantum-mechanical dynamical laws: Suppose that a certain system, subject to certain specified forces and constraints and whose state vector at time t_1 is $|A\rangle$, evolves, in accordance with the laws, into the state $|A'\rangle$ at time t_2; and suppose that that same system, subject to those same forces and constraints, if its state vector at t_1 is, rather, $|B\rangle$, evolves, in accordance with those laws, into the state $|B'\rangle$ at time t_2. Then, the laws dictate that if that same system, subject to those same forces and constraints, were, rather, in the state $\alpha|A\rangle + \beta|B\rangle$ at time t_1, then its state at time t_2 will be $\alpha|A'\rangle + \beta|B'\rangle$ (where $|A\rangle$ and $|B\rangle$ can be any state vectors at all). This property of the laws will concern us a good deal later on. The name of this property is *linearity* (and note that there is indeed a resemblance between "linearity" as applied to dynamical laws, here, and "linearity" as applied to operators, as in the two equations in (2.9)).

(D) The Connection with Experiment. So far, almost nothing in these principles has touched upon the results of measurements. All we have is a stipulation in (B) that the physical state whose state vector is an eigenvector, with eigenvalue *a*, of the operator associated with some particular measurable property will have the value *a* for that property; and presumably it follows that a measurement of that property, carried out on a system which happens to *be* in that state, will produce the result *a*. But much more needs to be said about the results of measurements than that! What if we measure a certain property of a certain physical system at a moment

when (as must happen in the vast majority of cases) the state vector of that system does not happen to be an eigenvector of that property operator? What if, say, we measure the color of a hard electron, an electron in a superposition of being white and being black? What happens then? Principle (B) is of no help here. A new principle shall have to be introduced to settle the question, which runs as follows:

Suppose we have before us a system whose state vector is $|a\rangle$, and we carry out a measurement of the value of property B on that system, where the eigenvectors of the property operator for B are $|B = b_i\rangle$, with eigenvalues b_i (i.e., $B|B = b_i\rangle = b_i|B = b_i\rangle$ for all i). According to quantum mechanics, the outcome of such a measurement is a matter of *probability*; and (more particularly) quantum mechanics stipulates that the probability that the outcome of this measurement will be $B = b_i$ is equal to:

(2.22) $(\langle a|B = b_i\rangle)^2$

Note that (as must be the case for probability) the number denoted by the above formula will always be less than or equal to 1; and note that in the special case of eigenvectors covered by principle (B), (2.22) yields (as it should) the probability 1. And note that it follows from (2.17) and (2.19) and (2.22) that the probability that a black electron will be found by a hardness measurement to be, say, soft, is (precisely as we have learned to expect) $\frac{1}{2}$.

And this is where it emerges that the correspondence between states and vectors of length 1 isn't precisely one-to-one. First of all, it follows from equation (2.3) that (for any vectors $|a\rangle$ and $|b\rangle$ and any number @) $\langle a|@|b\rangle = @\langle a|b\rangle$. Now, since the probability (2.22) depends only on the *square* of the product of the vectors involved, and since $(1x)^2 = (-1x)^2$, it follows that the probability of any result of any measurement carried out on a system in the state $|a\rangle$ will be identical to the probability of that same result of that same measurement carried out on a system in the state $-|a\rangle$. Vectors $|a\rangle$ and $-|a\rangle$, then, have precisely the same observable consequences; which is to say (as is customary in the quantum-mechanical literature) that the vectors $|a\rangle$ and $-|a\rangle$ represent precisely the same physical state.

(E) Collapse. Measurements (as I remarked in Chapter 1) are always, in principle, repeatable. Once a measurement is carried out and a result is obtained, the state of the measured system must be such as to guarantee that if that measurement is repeated, the same result will be obtained.[4]

Consider what that entails about the state vector of the measured system. Something happens to that state vector when the measurement occurs. If, say, a measurement of an observable called O is carried out on a system called S, and if the outcome of that measurement is $O = @$, then, whatever the state vector of S was just prior to the measurement of O, *the state vector of S just after that measurement must necessarily be an eigenvector of O* with eigenvalue @. The effect of measuring an observable must necessarily be to *change* the state vector of the measured system, to "collapse" it, to make it "jump" from whatever it may have been just prior to the measurement into some eigenvector of the measured observable operator. Which particular such eigenvector it gets changed into is of course determined by the outcome of the measurement; and note that that outcome, in accordance with principle (D), is a matter of probability. It's at this point, then, and at no point other than this one, that an element of pure chance enters into the evolution of the state vector.

Those are the principles of quantum mechanics. They are the most precise mechanism for predicting the outcomes of experiments on physical systems ever devised. No exceptions to them have ever been discovered. Nobody expects any.

Suppose that we should like to predict the behavior of some particular physical system by means of this algorithm. How, exactly,

4. Supposing, of course, that there has been no "tampering" in the interim; and supposing that not enough time has elapsed for the natural dynamics of the measured system itself to bring about changes in the value of the measured observable.

do we go about that? The first thing to do is to identify the vector space associated with that system: the space wherein all the possible physical states of that system can be represented. Given a precise physical description of the system, there are systematic techniques for doing that. Then the operators associated with the various measurable properties of that system need to be identified. There are techniques for doing that too. With that done, the specific correspondences between individual physical states and individual vectors can be mapped out (the vector which corresponds to the state wherein a certain measurable property has a certain value, for example, will be the one which is an eigenvector, with that eigenvalue, of the operator associated with that property). Then the present state vector of the system can be ascertained by means of measurements, and then (given the various forces and constraints to which the system will be subject) the state vector of any future time can be calculated by means of the prescription of principle (C), and then the probabilities of particular outcomes of a measurement carried out at some such future time can be calculated by means of principle (D), and the *effect* of such a measurement on the state vector can be taken into account by means of principle (E). And then principle (C) can be applied yet again, to that *new* state vector (the state vector which emerges from the measurement) to calculate the state vector of this system yet farther in the future, up to the moment when the next measurement occurs, whereupon principles (D) and (E) can be reapplied, and so on.

Notice, by the way, that principle (E) stipulates that under certain particular circumstances (namely, when a measurement occurs) the state vector evolves in a certain particular way (it "collapses" onto an eigenvector of the measured observable operator). Notice, too, that principle (C) is supposed to be a completely *general* account of how the state vector evolves under *any* circumstances. If that's all so, a question of consistency necessarily arises: it seems like (E) ought to be just a special case of (C), that (E) ought to be deducible from (C). But it isn't easy to see how that could be so, since the changes in the state vector stipulated by (E) are probabilistic, whereas those stipulated by (C) are, invariably, deterministic. This is going to require some worrying about, but let's not start that just yet; that worrying will commence in earnest in Chapter 4.

As I mentioned before, there is a standard way of talking, which students of physics are traditionally required to master along with this algorithm, about what superpositions are. That line, that way of dealing with the apparent contradiction of Chapter 1, boils down to this: the right way to think about superpositions of, say, being black and being white is to think of them as situations wherein color predicates cannot be applied, situations wherein color talk is unintelligible. Talking and inquiring about the color of an electron in such circumstances is (on this view) like talking or inquiring about, say, whether or not the number 5 is still a bachelor. On this view, then, the contradictions of Chapter 1 go away. On this view, it just isn't so that hard electrons are not black and not white and not both and not neither, since color talk of any kind, about hard electrons, simply has no meaning at all. And that's the way things are, on this view, for all sorts of superposition: superpositions are situations wherein the superposed predicates just don't apply.

Of course, once an electron has been *measured* to be white or black, then it *is* white or black (then, in other words, color predicates surely do apply). Measuring the color of a hard electron, then, isn't a matter of ascertaining what the color of that hard electron is; rather, it is a matter of first changing the state of the measured electron into one to which the color predicate applies, and to which the hardness predicate cannot apply (this is the "collapse" of principle (E)), and *then* of ascertaining the color of that newly created, color-applicable state. Measurements in quantum mechanics (and particularly within this interpretation of quantum mechanics) are very active processes. They aren't processes of merely learning something; they are invariably processes which drastically change the measured system.

That's what's at the heart of the standard view. The rest (of which I shall have much more to say later on) is details.

*   *   *

Here (before we move on to particular cases) are a few more general technicalities.

First, the vector spaces which are made use of in quantum mechanics are *complex* vector spaces. A complex vector space is one in which it's permissible to multiply vectors not merely by real numbers but by *complex* (i.e., real or imaginary or both) numbers in order to produce new vectors. In complex vector spaces, the expansion coefficients of vectors in given bases (the b_i of equation (2.1)) may be complex numbers too. That will necessitate a few refinements of what's been introduced thus far.

In complex vector spaces, the formula for the product of two vectors, written in terms of their expansion coefficients in some particular basis (that is, the formula (2.4b)), needs to be changed, very slightly (what, precisely, it gets changed into need not concern us here), in order to guarantee that the norm of any vector (that is, its length: $\sqrt{\langle A|A \rangle}$) remains, under all circumstances, a positive real number. Formula (2.22) for probabilities needs to be altered very slightly too, since, in complex spaces, $\langle A|B \rangle$ and, hence, (2.22) may be complex numbers (and yet probabilities must necessarily be real, positive numbers between 0 and 1). The solution is to change (2.22) to

$$(2.23) \qquad |\langle a|B = b_i \rangle|^2$$

where the vertical bars denote absolute value (or "distance from zero," which is invariably a real, positive number). Equation (2.23) stipulates that the probability that a measurement of B on a system in the state $|a\rangle$ will produce the outcome $B = b_i$ is equal to the square of the distance from 0 of the complex number $\langle a|B = b_i \rangle$; and that probability, so defined, will invariably be a real and positive number. Formula (2.22), by the way, will entail not only that $|A\rangle$ and $-|A\rangle$ represent the same physical state (we've already seen that to be the case), but, more generally, that $|A\rangle$ and $@|A\rangle$ represent the same state, where @ may be any one of the infinity of complex numbers of absolute value 1.

The elements of the operator matrices of linear operators on complex vector spaces (that is, the numbers O_{ij} of (2.10) and (2.11)) can be complex numbers too. Nonetheless, it may happen to some such operators that all of their eigenvectors are associated only with real eigenvalues (albeit, perhaps, not all of their matrix elements O_{ij}, and perhaps even none of them, are real). Linear operators like that are called *Hermitian* operators; and it's clear from principle (B) (since, of course, the values of physically measurable quantities are always real numbers) that the operators associated with measurable properties must necessarily be Hermitian operators.

Here are some facts about Hermitian operators:

(1) If two vectors are both eigenvectors of the same Hermitian operator, and if the eigenvalues associated with those two eigenvectors are two different (real) numbers, then the two vectors in question are necessarily orthogonal to each other.

That pretty much had to be so, if this algorithm is going to work out right; otherwise, measurements wouldn't be repeatable. The different eigenvalues of a property operator, after all, correspond to different values of that property; and (if measurements of a property are to be repeatable) having a certain value of a certain property must entail that subsequent measurements of that property will certainly not find any other value of it;[5] and that (given principle (D)) will require that state vectors connected with different values of the same measurable property ($|$black$\rangle$ and $|$white$\rangle$, say, or $|$hard$\rangle$ and $|$soft$\rangle$) be orthogonal to one another.

(2) Any Hermitian operator on an N-dimensional space will always have at least one set of N mutually orthogonal eigenvectors. Which is to say: it will always be possible to form a basis of the space out of the eigenvectors of any Hermitian operator; different bases, of course, for different operators. Consider, for example, the hardness operator of equation (2.18) and the color operator of equation (2.19).

(3) The reader ought to be able to persuade herself, now, of the following: if a Hermitian operator on an N-dimensional space

5. Supposing, once again, that no tampering, and no dynamical evolution, has gone on in the meantime.

happens to have N different eigenvalues, then there is a *unique* vector in the space (or, rather, unique modulo multiplication by numbers) associated with each different one of those eigenvalues; and of course the set of all eigenvectors of length 1 of that operator will form a unique basis of that space (or, rather, unique modulo multiplication by numbers of absolute value 1). Operators like that are called *complete* or *nondegenerate* operators.

(4) Any Hermitian operator on a given space will invariably be associated with some measurable property of the physical system connected with that space (this is just a somewhat more informative version of the first part of principle (B)).

(5) Any vector whatever in a given space will invariably be an eigenvector of some complete Hermitian operator on that space. That, combined with fact (4) and principle (B), will entail that any quantum state whatever of a given physical system will invariably be associated with some definite value of some measurable property of that system.

All this turns out to entail (among other things) that every quantum-mechanical system necessarily has an infinity of mutually incompatible measurable properties. Think (just to have something concrete to talk about) of the space of possible spin states of an electron. There are, to begin with, a continuous infinity of different such states (since there are a continuous infinity of vectors of length 1 in a two-dimensional space); moreover, given any one of those states, there are clearly a continuous infinity of different possible states which are not orthogonal to it. And, by facts (3) and (5) above, every state in this space is necessarily the only eigenstate associated with a certain particular eigenvalue of a certain particular complete operator, and, by fact (1), none of the continuous infinity of states which aren't orthogonal to the state in question can possibly be eigenstates of the same complete operator. What's more, the complete operators of which those other states *are* eigenstates clearly can't even be *compatible* with the operator in question. And so (since all this applies to every state in the space) there must necessarily be a continuous infinity of mutually incompatible complete measurable properties, of which color and hardness are only two.

It will be useful, for what comes later, to give two more of those properties names. The vectors

$$1/2|\text{black}\rangle + \sqrt{3}/2|\text{white}\rangle \quad \text{and} \quad \sqrt{3}/2|\text{black}\rangle - 1/2|\text{white}\rangle$$

are both of length 1 and are orthogonal to one another (and aren't orthogonal to any of the eigenvectors of color or hardness), and so it follows that there must be a complete observable of which they are both eigenstates, with different eigenvalues (which can always be set at $+1$ and -1, respectively). Let's call that observable "gleb." And the vectors

$$1/2|\text{black}\rangle - \sqrt{3}/2|\text{white}\rangle \quad \text{and} \quad \sqrt{3}/2|\text{black}\rangle + 1/2|\text{white}\rangle$$

are both of length 1 and are orthogonal to one another (and aren't orthogonal to any of the eigenvectors of color or hardness or gleb), and so it follows that there must be a complete observable of which *they* are both eigenstates, with different eigenvalues (which can always be set at $+1$ and -1, respectively). Let's call that observable "scrad." Of course, the eigenstates of gleb and scrad (just like those of color and hardness) both form different bases of the spin space.

Finally, there are rules (never mind what those rules are, precisely) for adding and subtracting matrices to or from one another, and for multiplying them by one another. The *commutator* of two matrices A and B, which is denoted by the symbol $[A,B]$, is defined to be the object $AB - BA$ (the rules for multiplying matrices by one another entail that the order of multiplication counts: AB isn't necessarily the same as BA).

Now, it can be shown that in the event that $[A,B] = 0$ (that is, in the event that AB *is* equal to BA), A and B *share* at least one set of eigenvectors which form a basis of the space. A little reflection will confirm that the operator matrices of incompatible observables can't possibly share any such complete basis of eigenvectors (since such eigenvectors would correspond to definite value states of both observables at the same time). It must be the case, then, that *the commutators of incompatible observable matrices are nonzero.* So the property of commutativity (that is, the condition $[A,B] = 0$) turns out to be a convenient mathematical test for compatibility. Moreover, in cases of incompatible observables, the commutator of the two observables in question turns out to be extremely useful for assessing the *degree* of their incompatibility.[6]

Coordinate Space

Let's begin to apply all this. Let's see, in some detail, how to set up a quantum-mechanical representation, and a quantum-mechanical dynamics, of some simple physical system. Forget about color and hardness for the moment. Think of a familiar sort of particle, one with only the familiar sorts of physical properties: position and velocity and momentum and energy and things like that.

Here's a way to get started: We know, from hundreds of years of experience, that the behaviors of relatively big particles, with relatively big masses (particles you can see, like rocks and baseballs and planets) are very well described by the classical mechanics of Newton. That entails something about the quantum theory of particles: whatever that theory ends up predicting about the strange, tiny particles of Chapter 1, it ought to predict that everyday particles, subject to everyday circumstances, will behave in the

6. Perhaps the notion of there being various different degrees of incompatibility requires some elucidation. Here's what the idea is (or here's what it is, at any rate, in the simplest case, when the observables involved are both complete):

Consider two complete and incompatible observables (call them A and B) of some physical system. If, when any particular eigenstate of A obtains, the outcome of a measurement of B can be predicted (by means of formula (2.23)) with something approaching certainty (that is: if, for each eigenvector of A, there is some particular eigenvector of B such that the product of those two vectors is something approaching one), then A and B are said to be only very slightly incompatible. But if (at the other extreme), when any particular eigenstate of A obtains, the probabilities of the various possible outcomes of a measurement of B are all the same (that is: if knowing the value of A gives us no information whatever about the outcome of an upcoming measurement of B), *then* A and B are said to be *maximally* incompatible.

So (for example) color and hardness (which are maximally incompatible observables) are a good deal more incompatible with one another than color and scrad are.

THE MATHEMATICAL FORMALISM

everyday, Newtonian way. It turns out that that requirement (which is referred to in the quantum-mechanical literature as the principle of *correspondence*) can be parlayed into a prescription for calculating the commutators of the quantum observable operators from mathematical relations among the corresponding measurable properties of the classical theory.[7]

Now, it happens that this prescription implies that the momentum and the position of a particle are incompatible observables. The commutator of p and x (p is the traditional symbol for the momentum of a particle, and x stands for position) is

(2.24) $[p,x] = i$

where is a number, a physical constant, called Planck's constant, and i is the imaginary number $\sqrt{-1}$. The important thing about (2.24), of course, is just that it isn't zero.

So, there will be a basis of the space of possible states of such particles consisting entirely of eigenvectors of the x operator, and there will be a basis of that space consisting entirely of eigenvectors of the p operator, and (since x and p are incompatible) those two bases won't consist of the same vectors. A state of some definite momentum will be a superposition of various different states of definite position, and a state of some definite position will be a superposition of various different states of definite momentum (just as it happened with color and hardness).[8]

Let's start to explore this space. Let's look at the x basis (the basis

7. Actually, the logical relationship between that latter prescription (which relates the Poisson brackets of the classical theory to the commutators of the quantum theory) and the correspondence principle has long been a subject of dispute. The prescription's ultimate justification is that it seems to work.

8. What about the consistency of all that with the known behaviors of, say, baseballs? Baseballs, after all, *do* have quite definite positions *and* quite definite velocities at the same time! And velocities, after all, are just momenta divided by the mass of the particle. Here's how that works: Since velocities *are* just momenta divided by mass, ranges (uncertainties) of velocities are just ranges of momenta divided by mass. So a big mass means a small velocity-uncertainty (even when the momentum-uncertainty is large). So baseballs (whose masses are relatively gigantic) can behave just as they always do and yet be fully in accordance with the laws of quantum mechanics.

of position eigenvectors). Let's call the position operator X. Consider a particle which is confined (to keep things simple for the moment) to a one-dimensional coordinate space; a particle which is constrained to move along a line. Let $|X = 5\rangle$ represent the state in which that particle is located at the point 5. Then, in accordance with principle (B):

$$(2.25) \quad X|X = 5\rangle = 5|X = 5\rangle$$

Note that the possible eigenvalues of X (unlike those of color and hardness) will form a continuum extending from $-\infty$ to $+\infty$ (since the points on a line, the possible locations of such a particle, are continuous and infinitely extended).

Now, since (in accordance with the facts about Hermitian operators that were recited at the end of the last section) the various different eigenvectors of X must necessarily form a basis of the state space of this particle, and since X has an infinity of different eigenvalues, and since the eigenvectors $|X = @\rangle$ associated with those different eigenvalues must necessarily all be orthogonal to one another, it follows that the state space of this particle must necessarily be infinite-dimensional! And here, by the way, a particularly dangerous confusion is to be scrupulously avoided. There are two "spaces" coming into play here: the one-dimensional coordinate space, which is the space of locations, the familiar, ordinary, physical space in which the particle is free to move around; and the much more abstract vector space of states, which is here infinite-dimensional and of which the locations (which constitute the entire coordinate space) merely form a basis. The two shouldn't get mixed up.

The fact that the X eigenvectors form a basis of the state space also entails that any vector whatever ($|\psi\rangle$, say) in that infinite-dimensional space can be expanded (in accordance with (2.1) and (2.2)) in terms of X eigenstates like this:

$$(2.26) \quad |\psi\rangle = a_5|X = 5\rangle + a_7|X = 7\rangle + a_{72.93}|X = 72.93\rangle + \ldots$$
$$\text{where } a_x = \langle\psi|X = x\rangle$$

Let me introduce a notation now which will serve precisely the same purpose for infinite-dimensional vector spaces as the column-vector notation of equation (2.5) serves for finite-dimensional vector spaces. Think of a_x in (2.26) as a function of x:

$$(2.27) \qquad a_x = \langle \psi | X = x \rangle = \psi(x)$$

Just as the N numbers in (2.5), in a given basis, pick out a unique vector, the function $\psi(x)$ (the infinite list of correspondences between a values and x values), in the X basis which is implicit here, serves to pick the unique vector $|\psi\rangle$ out of the infinite-dimensional space. Given a basis choice (which, as I said, is implicit here), a ψ function (a *wave function*, as it's called in the literature) carries precisely as much information as does $|\psi\rangle$ itself. As a matter of fact, $\psi(x)$ constitutes a blueprint from which (a la (2.26)) $|\psi\rangle$ can be explicitly constructed, and vice versa.

It follows from (2.26), for example, that the function $\psi(x)$ for the state $|X = 5\rangle$ is the function with value 1 at the point $x = 5$ and 0 elsewhere and that the function $\psi(x)$ for the state $1/\sqrt{2}|X = 3\rangle - 1/\sqrt{2}|X = 7\rangle$ is the function with value $1/\sqrt{2}$ at the point $x = 3$ and the value $-1/\sqrt{2}$ at the point $x = 7$, and 0 elsewhere.

Location probabilities can be read off from the wave functions too. It follows from (2.27) and (2.22) that if the wave function of a certain particle at a certain time is $\psi(x)$, and if a position measurement is carried out at that time on that particle, then the probability that that measurement will find that particle to be located at the point $x = x_1$, say, will be equal to $|\psi(x_1)|^2$ (that is: the square of the magnitude of $\psi(x)$ at the point x_1).[9]

Moreover, any measurable property of particles (momentum,

9. All this is a little bit oversimplified. The way I've been talking over the last few paragraphs (and the way I'll be talking throughout the rest of this book, except on special occasions) is as if the points in space which an electron can potentially occupy form a discrete set. But what those points really form, of course, is a continuum. And so sums like the one in (2.26) really ought to be integrals; and the sorts of position probabilities one typically calculates in quantum mechanics are really probabilities of certain particles being found in certain finite regions, and not probabilities of finding them at certain particular points.

energy, whatever) turns out to be representable as an operator on the wave function (namely, as some prescription for taking one wave function into another), rather than on the state vectors; and eigenfunction-eigenvalue-operator relations precisely analogous to those stipulated in equation (2.14) and in principle (B) apply here; and there are rules for adding and multiplying wave functions (analogous to (2.4a) and (2.4b)) whereby the sums and products of state vectors can be calculated; and even the equations of motion of the state vector can be recast as equations of motion for the wave function. To make a long story short, anything whatever that can be said about the state vectors of particles can be translated into the language of wave functions. And that's pretty much all there is to the quantum mechanics of single, structureless particles.

Systems Consisting of More Than a Single Particle

We shall need to know something about the quantum mechanics of systems consisting of more than one particle too. A two-particle system will suffice as an example. Let's set up the state space of a system like that.

Here's how to start: Imagine a pair of particles, one of which (number 1) is in the state $|\psi_a\rangle$ and the other of which (number 2) is in the state $|\psi_b\rangle$. The quantum state vector of a pair of particles like that is traditionally written down like this:

$$(2.28) \qquad |\psi_a\rangle_1 |\psi_b\rangle_2$$

or like this:

$$(2.29) \qquad |\psi_a^1, \psi_b^2\rangle$$

These are taken to represent a vector in the state space of the two-particle system.

Let's reason out some of the properties of such systems. The

But none of that is going to make any difference whatever to the sorts of questions I want to talk about here.

probability calculus, to begin with, is going to get more complicated: we shall have to deal, now, with *joint* probabilities involving
the outcomes of one sort of measurement carried out on, say,
particle 1 and the outcomes of another sort of measurement carried
out, at the same time, on particle 2. Suppose that the two particles
described above don't interact with one another. Then the familiar
laws of composition for independent probabilities ought to apply:
The probability that the outcome of a measurement of property *A*
on particle 1 is *A* = *a and* the outcome of a measurement of
property *B* on particle 2 is *B* = *b* ought to be equal to the probability of the former outcome *times* the probability of the latter one.
Those two probabilities, of course, are ones that we know how to
calculate from the one-particle theory. Now, if the two-particle
probability calculations are to proceed in accordance with (2.22),
and if the laws of composition are to apply, then it's easy to show
that the rule for multiplying vectors like (2.28) or (2.29) by one
another is going to have to be:

(2.30) $\langle \psi_c^1, \psi_d^2 | \psi_a^1, \psi_b^2 \rangle = \langle \psi_c^1 | \psi_a^1 \rangle \times \langle \psi_d^2 | \psi_b^2 \rangle$

Let's go on.[10] Suppose that the vectors $|\psi_1^1\rangle \ldots |\psi_N^1\rangle$ constitute a
basis of the state space of particle 1, and that $|\psi_1^2\rangle \ldots |\psi_N^2\rangle$ consti-
tute a basis of the state space of particle 2 (of course, we've just
learned that each of those bases must in fact consist of a continuous
infinity of vectors; but pretend, for the moment, to make things
simpler, that that isn't so). Then (2.30) entails that

(2.31) $\langle \psi_i^1, \psi_j^2 | \psi_k^1, \psi_l^2 \rangle = 0$ unless $i = k$ and $j = l$

10. We're doing things kind of backwards here, of course. If we knew precisely
what sort of vector space we were dealing with, and if we knew precisely what
sorts of vectors are represented by (2.28) and (2.29), then (2.30) could simply be
derived from, say, (2.4b), and then the composition law for probabilities would
follow from (2.30) and (2.22). As things stand, however, we *don't* know what sorts
of vectors (2.28) and (2.29) represent. Our strategy for finding that out is to *assume*
the composition law, to derive (2.30) from *that* (together with (2.22)), and then to
use (2.30) to ascertain what sorts of vectors and vector spaces we're dealing with.

and *that* entails that the dimensionality of the two-particle state space will be equal to the dimensionality of the state space of particle 1 *times* the dimensionality of the state space of particle 2. The dimensionality of the two-particle space, in other words, is N^2. The entire set of N^2 vectors of the form

(2.32) $|\psi_i^1, \psi_j^2\rangle$ for $i, j = 1 \ldots N$

will all be orthogonal to one another, and they will form a *basis* of the two-particle space; and (as always) any linear combination of those basis vectors of the form

(2.33) $|\psi^{1,2}\rangle = a_{11}|\psi_1^1, \psi_1^2\rangle + a_{12}|\psi_1^1, \psi_2^2\rangle + \ldots$

will be another vector, another possible state, in that space.

Now something interesting comes up. Consider a state of the two-particle space of the form:

(2.34) $|Q\rangle = 1/\sqrt{2}|\psi_1^1, \psi_1^2\rangle + 1/\sqrt{2}|\psi_2^1, \psi_2^2\rangle$

It can be shown that the state $|Q\rangle$ cannot possibly be recast (no matter what bases we choose within the spaces of the two individual particles) in the form $|f^1, g^2\rangle$. That is, states like (2.34) cannot possibly be decomposed into a well-defined state of particle 1 and a well-defined state of particle 2; states like that cannot possibly be described by propositions of the form "the state of particle 1 is such-and-such" and "the state of particle 2 is such-and-such." In states like (2.34) (and this is just another way of saying the same thing), no measurable property of particle 1 alone, and no measurable property of particle 2 alone, has any definite value. The smallest system to which any state can be assigned here, the smallest system which can be assigned a definite value of any measurable property,[11] is the two-particle system. States like that are called *nonseparable;* and the phenomenon of nonseparability (like those

11. We're not talking about properties like mass or charge here, of course (those are simply among the *defining* characteristics of particles), but about coordinate-space properties like position or momentum or velocity or energy.

THE MATHEMATICAL FORMALISM

of superposition and incompatibility) is widely thought to be one of the most profound differences there is between quantum mechanics and the classical picture of the world.

Here's an example: Consider the state

$$(2.35) \quad |Q'\rangle = \frac{1}{\sqrt{2}}|X^1 = 5\rangle|X^2 = 7\rangle + \frac{1}{\sqrt{2}}|X^1 = 9\rangle|X^2 = 11\rangle$$

which I've written down in the notation of (2.28) (X^1 and X^2 are the X operators of particles 1 and 2, respectively). Neither the position of particle 1, nor the position of particle 2, nor anything else about either of them separately, has any definite value here; but the difference in their positions does have a definite value;[12] that is:

$$(2.36) \quad (X^2 - X^1)|Q\rangle = 2$$

Let me say something more about the calculations of probabilities of experimental results in the two-particle case. Formula (2.22) is, of course, still the rule; but the application of (2.22) to two-particle systems will require some elaborating. There are two interesting cases.

Suppose, first, that the state of the two-particle system is $|k\rangle$, and that A^1 and B^2 (which are complete observables of particles 1 and 2, respectively) are measured (the example discussed just above equation (2.30) is a case like that; in that case, the state $|k\rangle$ happens to be separable). The probability that the outcomes of those experiments will be $A^1 = a_i$ and $B^2 = b_i$ is

$$(2.37) \quad |\langle A^1 = a_i, B^2 = b_i|k\rangle|^2$$

Now suppose that the two-particle state is $|k\rangle$ and that only A^1 is measured. The probability that the outcome of that measurement will be $A^1 = a_i$ is

$$(2.38) \quad |\langle A^1 = a_i, L^2 = l_1|k\rangle|^2 + |\langle A^1 = a_i, L^2 = l_2|k\rangle|^2 + \ldots$$

12. Perhaps it's worth saying explicitly, at this juncture, that the way one applies one-particle operators to two-particle states is exactly the way one would think one does: $|A^1 = w, B^2 = z\rangle$ is an eigenvector of A^1 with eigenvalue w, and of B^2 with eigenvalue z, and of $A^1 - B^2$ with eigenvalue $w - z$.

where L^2 is some complete observable of particle 2, and where the sum ranges over all the eigenvalues l_i of L. L can be any complete observable of particle 2 at all; no matter which one we pick, the answer will come out the same. The intuition connected with (2.38) is something like this: the probability that A^1 comes out to be a_i is equal to the sum of the probabilities of all of the various different ways in which it might come to pass that A^1 comes out to be a_i.

The principle of collapse for two-particle systems will need some elaborating too. Principle (E) is still the rule, but we shall need to say more precisely what (E) means for two-particle systems. Suppose, then, that the state of a certain two-particle system just prior to the time t_1 is $|D\rangle$; and suppose that at t_1 the observable A^1 (of particle 1) is measured, and suppose that the outcome of that measurement is $A^1 = a_5$. Here's how to calculate the state of the two-particle system just after t_1 (here, that is, is how to calculate what state the state of the two-particle system gets changed into by the A^1 measurement): Start with the state $|D\rangle$ expressed in terms of eigenstates of A^1 and L^2 (where L^2 is any complete observable of particle 2; no matter which one we pick, this calculation will come out the same):

$$(2.39) \quad |D\rangle = d_{11}|A^1 = a_1, L^2 = l_1\rangle + d_{12}|A^1 = a_1, L^2 = l_2\rangle + \ldots$$
$$+ d_{1N}|A^1 = a_1, L^2 = l_N\rangle + d_{21}|A^1 = a_2, L^2 = l_1\rangle + \ldots$$

where

$$(2.40) \quad d_{ij} = \langle A^1 = a_i, L^2 = l_j|D\rangle$$

Then, throw away all of the terms in (2.39) other than the ones for which $A^1 = a_5$. Then, multiply all the remaining d_{ij}'s (that is, the d_{5j}'s) by some number (the same number for all the d_{5j}'s, so as not to alter their relative magnitudes), so as to make the remaining part of (2.39) a vector of norm 1. And that new vector is the state vector of this two-particle system just after the measurement.

Here are some examples. For one-particle systems (for which, of course, there *is* no L^2 property) this generalized prescription for collapse will entail that the effects of measurements are just as they were described in principle (E). For two-particle systems in separable states, this principle will entail that the measurement only affects the state of the measured particle. Here's how that works: Suppose that the premeasurement state $|D\rangle$ is the separable state

$$(2.41) \quad |D\rangle = |Q^1 = w, M^2 = z\rangle$$

suppose that A^1 is measured, with the outcome $A^1 = a_5$. Then, let L^2 in (2.39) be M^2 (remember that we can choose any operator we like for L^2), and (2.39) will take the form:

$$(2.42) \quad |D\rangle = d_1|A^1 = a_1, M^2 = z\rangle + d_2|A^1 = a_2, M^2 = z\rangle + \ldots$$

with

$$(2.43) \quad d_i = \langle A^1 = a_i, M^2 = z|Q^1 = w, M^2 = z\rangle = \langle A^1 = a_i|Q^1 = w\rangle$$

And so, following the instructions below (2.40), we end up with the postmeasurement *(pm)* state:

$$(2.44) \quad |D\rangle_{pm} = |A^1 = a_5, M^2 = z\rangle$$

The interesting cases, though, are the nonseparable ones. In those cases (unlike in the separable ones) the measurement brings about changes in the quantum mechanical description of the *un*measured particle as well. Suppose, for example, that the premeasurement state is

$$(2.45) \quad |D\rangle = \tfrac{1}{\sqrt{2}}|A^1 = a_4, L^2 = l_7\rangle + \tfrac{1}{\sqrt{2}}|A^1 = a_5, L^2 = l_{24}\rangle$$

where, once again, A_1 is eventually measured, with the result $A^1 = a_5$. Formula (2.45) has already been written down in the form of (2.39), so we can proceed directly to carrying out the instructions below (2.40). Once those instructions have been carried out, we end up with:

$$(2.46) \quad |D\rangle_{pm} = |A^1 = a_5, L^2 = l_{24}\rangle$$

One more thing needs to be described before we can talk about the story in chapter 1. If the values of M separate and independent physical properties of a certain physical system need to be specified in order to uniquely pick out that system's physical state, then that system is said to have M *degrees of freedom*. A single, structureless particle confined to a one-dimensional coordinate space has 1 degree of freedom; and a system consisting of two such properties (such as we have just now been considering) has 2 degrees of freedom; and a single particle free to move in a three-dimensional coordinate space has 3; and a particle that has color-hardness properties and is free to move in a three-dimensional coordinate space has 4. The quantum-mechanical treatment of systems with multiple degrees of freedom is precisely analogous to that of multiple *particle* systems just described. The states of such systems, as for multiple-particle systems, are written down, degree of freedom by degree of freedom, side by side, as in (2.28) and (2.29). For example, the state of a white electron whose coordinate-space wave function is $\psi(x)$ is written (in the hardness basis of equation (2.17)) as

$$(2.47) \qquad \begin{bmatrix} 1/\sqrt{2} \\ -1/\sqrt{2} \end{bmatrix} |\psi\rangle$$

And everything I've said here about multiple-particle systems applies straightforwardly to multiple-degree-of-freedom systems too.

The Two–Path Experiments

Now we're ready to retell the crucial stories of Chapter 1. The third two-path experiment (the one that's so perplexing) is mapped out carefully, with the help of a coordinate system, in figure 2.8. The times at which the various different stages of the experiment unfold are indicated there too.

At time t_1, when the particle is about to enter the apparatus, its state is:

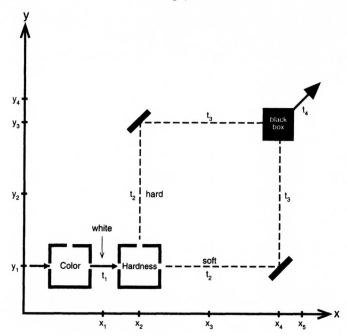

Figure 2.8

(2.48) $\quad |\text{white}, X = x_1, Y = y_1\rangle$

$$= \begin{bmatrix} 1/\sqrt{2} \\ -1/\sqrt{2} \end{bmatrix} |X = x_1, Y = y_1\rangle$$

$$= (1/\sqrt{2}\begin{bmatrix} 1 \\ 0 \end{bmatrix} - 1/\sqrt{2}\begin{bmatrix} 0 \\ 1 \end{bmatrix}) |X = x_1, Y = y_1\rangle$$

$$= \underbrace{1/\sqrt{2}|\text{hard}\rangle|X = x_1, Y = y_1\rangle}_{|a\rangle} - \underbrace{1/\sqrt{2}|\text{soft}\rangle|X = x_1, Y = y_1\rangle}_{|b\rangle}$$

where, as usual, I've written out the spin vectors in the hardness basis. Here's how to calculate what happens next. Consider this: If the state at t_1 weren't (2.48) but, rather, just $|a\rangle$, and if the hardness box really is a hardness box, then the state at time t_2 would be

(2.49) $\quad |\text{hard}\rangle|X = x_2, Y = y_2\rangle$

And if the state at t_1 were just $|b\rangle$, then the state at t_2 would be

(2.50) $|\text{soft}\rangle|X = x_3, Y = y_1\rangle$

However, as the state at t_1 is in fact neither $|a\rangle$ nor $|b\rangle$ but, rather, $1/\sqrt{2}|a\rangle - 1/\sqrt{2}|b\rangle$, it follows from (2.49) and (2.50) and from the linearity of the quantum dynamics (which was spelled out in principle (D)) that the state at t_2 is really

(2.51) $1/\sqrt{2}|\text{hard}\rangle|X = x_2, Y = y_2\rangle - 1/\sqrt{2}|\text{soft}\rangle|X = x_3, Y = y_1\rangle$

This state, by the way, involves nonseparable correlations between spin and coordinate-space properties of the electron: No spin property of the electron in this state (neither hardness nor color nor anything else) nor any of its coordinate-space properties (position, momentum, etc.) has any definite value here, just as no property of either particle 1 or particle 2 is separately definite in the state of equation (2.34). The only properties which *are* definite in (2.51) involve combinations of spin-space *and* coordinate-space variables of the particle. We'll talk about those later on.

Formula (2.51) represents a superposition of states, in one of which the electron is traveling along the hard path and in the other of which the electron is traveling along the soft path. It is of the state in formula (2.51) that we were compelled to conclude, in Chapter 1, that it's false that the electron takes the hard path, and false that it takes the soft one, and false that it takes both, and false that it takes neither; and the problem was that those four claims together amount to a contradiction. On the standard way of thinking, as I've mentioned already, those hypotheses (hard path, soft path, both, neither) aren't false, they're meaningless, they're categorical mistakes.

Let's go on. The same sort of reasoning as led from (2.48) to (2.51) will imply, starting from (2.51), that the state of the electron at t_3 is:

(2.52) $1/\sqrt{2}|\text{hard}\rangle|X = x_3, Y = y_3\rangle - 1/\sqrt{2}|\text{soft}\rangle|X = x_4, Y = y_2\rangle$

and the same sort of reasoning applied, in turn, to (2.52) will imply that the state at time t_4 is:

(2.53) $\quad 1/\sqrt{2}|\text{hard}\rangle|X = x_5, Y = y_4\rangle - 1/\sqrt{2}|\text{soft}\rangle|X = x_5, Y = y_4\rangle$

$\qquad = 1/\sqrt{2}(|\text{hard}\rangle - |\text{soft}\rangle)|X = x_5, Y = y_4\rangle$

$\qquad = |\text{white}\rangle|X = x_5, Y = y_4\rangle$

At this point, the spin state and the coordinate-space state have become separable again. The position of the electron once again has a definite value now, and that (as can be seen from the calculation in (2.53)) renders its color definite too. So, the fact that a hard electron fed into this device will come out hard, and that a soft electron will come out (at that same point) soft, together with the linearity of the quantum dynamics, entails that a white electron fed into this device will come out (just as we found it did) white.

What if we were to stop the experiment in the middle by measuring the position of the electron at, say, t_3? Then the superposition would go away; a collapse (much like the collapse described in equations (2.45)–(2.46)) would occur, and the state just after the measurement would be either

(2.54) $\quad |\text{hard}\rangle|X = x_3, Y = y_3\rangle \quad$ or $\quad |\text{soft}\rangle|X = x_4, Y = y_2\rangle$

each with a probability of 1/2 (in accordance with (2.22) and (2.38)), and the state at t_4 would be (respectively)

(2.55) $\quad |\text{hard}\rangle|X = x_5, Y = y_4\rangle \quad$ or $\quad |\text{soft}\rangle|X = x_5, Y = y_4\rangle$

What if we put a wall into the soft path at y_1, x_3? Then the state at t_4 would be:

(2.56) $\quad 1/\sqrt{2}|\text{hard}\rangle|X = x_5, Y = y_4\rangle - 1/\sqrt{2}|\text{soft}\rangle|X = x_3, Y = y_1\rangle$

In this instance, then, the state remains nonseparable between the spin and coordinate-space properties at t_4. If a measurement of the position of this electron were to be carried out at t_4 (if, say, we were to look and see whether the electron had emerged from the black box), the probability of finding it at (x_5, y_4) would be ½, and if it were found there it would be hard, and if its color were measured, it would be as likely to be found black as white; and all that, once again, is precisely in accord with the results of the experiments described in Chapter 1.

The state described in (2.56), as I just mentioned, is nonseparable between spin and coordinate space, so it isn't associated with any definite values for hardness or color or position or momentum or anything like that. Nonetheless, (2.56) *is* a quantum state (that is, it's a vector in the state space of this electron), so (in virtue of fact (5) about Hermitian operators) it must be associated with definite values of *some* measurable properties.

Let's find out what those properties are. First let's simplify our notation a bit. The electron in (2.56) has only two possible positions, so let's replace the position operator (which has an infinity of possible eigenvalues, most of which are irrelevant here) with something we'll call a "where" operator (which has only two possible eigenvalues). "Where = +1" means $X = x_5$, $Y = y_4$; "where = −1" means $X = x_3$, $Y = y_1$. If we represent the eigenvectors of where like this:

(2.57)
$$|X = x_5, Y = y_4\rangle = \begin{bmatrix} 1 \\ 0 \end{bmatrix} \qquad |X = x_3, Y = y_1\rangle = \begin{bmatrix} 0 \\ 1 \end{bmatrix}$$

then

(2.58)
$$\text{where} = \begin{bmatrix} 1 & 0 \\ 0 & -1 \end{bmatrix}$$

So, (2.56) can be rewritten like this:

(2.59)
$$1/\sqrt{2} \begin{bmatrix} 1 \\ 0 \end{bmatrix}_{\text{hard}} \begin{bmatrix} 1 \\ 0 \end{bmatrix}_{\text{where}} - 1/\sqrt{2} \begin{bmatrix} 0 \\ 1 \end{bmatrix}_{\text{hard}} \begin{bmatrix} 0 \\ 1 \end{bmatrix}_{\text{where}}$$

The subscripts to the column vectors above indicate what degrees of freedom those vectors refer to. Now let's define one more Hermitian operator (one more observable, that is); one which operates on the position space. Let's call it "zap":

$$\text{(2.60)} \qquad \text{zap} = \begin{bmatrix} 0 & 1 \\ 1 & 0 \end{bmatrix}$$

written in the where basis. The reader can now confirm for herself that (2.59) is an eigenstate of hard minus where, with eigenvalue 0, and that it's an eigenstate of color minus zap, with eigenvalue 0; and it turns out that those two eigenvalues uniquely pick out the state (2.59). Precisely what sort of measurable property zap is, and precisely how to measure it, is a complicated matter; but (in accordance with fact (4) about Hermitian operators) it is, with certainty, such a property.

Let me, finally, say something about how to build the sort of total-of-nothing box described in Chapter 1. Consider a system whose state vector, at a certain moment, is $|A\rangle$; and suppose that a measurement of a complete observable B is to be carried out on that system. The probabilities of the various possible outcomes of that measurement are given by equation (2.23). Now, suppose that the state of that system were not $|A\rangle$ but, rather, $-|A\rangle$. Since (2.23) depends not on $\langle A|B = b_i\rangle$ itself but, rather, only on the *square* of $\langle A|B = b_i\rangle$, the change from $|A\rangle$ to $-|A\rangle$ won't make any difference in any of those probabilities, no matter what observable B may happen to be. Suppose, for example that

$$\text{(2.61)} \qquad |A\rangle = |\text{hard}\rangle = \begin{bmatrix} 1 \\ 0 \end{bmatrix}$$

Then

$$\text{(2.62)} \qquad -|A\rangle = -1 \times \begin{bmatrix} 1 \\ 0 \end{bmatrix} = \begin{bmatrix} -1 \\ 0 \end{bmatrix}$$

and it follows from (2.4b) that the square of the product of any column vector with (2.62) will be equal to the square of the product of that same vector (whatever vector that is) with (2.61).

So, a box which changes the state of any incoming electron into -1 times that incoming state will be a total-of-nothing box, since it will change none of the values, nor any probabilities of values, of any of the observables of any electron which passes through it; needless to say, such a box will have no effect whatever on the state of an electron which passes outside of it.

But the effects of such a box on an electron which is in a superposition of passing through it and outside of it may be quite a different matter. Suppose, for example, that such a box is inserted in the soft path of our two-path device at $(x_{3.5}, y_1)$. Then, if the initially input electron was white, the state at t_2 will be, as above, (2.51), and the state at t_3 (*after* the "passage" through the box) will be not (2.52) but, rather,

$$(2.63) \qquad 1/\sqrt{2}|\text{hard}\rangle|X = x_3, Y = y_3\rangle + 1/\sqrt{2}|\text{soft}\rangle|X = x_4, Y = y_2\rangle$$

(The sign of the second term is changed, relative to (2.52), by the passage through the box.) The same reasoning as led from (2.52) to (2.53) will now imply that the color of the electron at t_4 is, with certainty, black.

Field Theory

There's just one more thing that it will turn out to be convenient (a good deal later on) to have mentioned here.

It has to do with adjusting the quantum-mechanical formalism so as make it consistent with special relativity.

It turns out that this adjustment doesn't require any change at all in principles (A)–(D). What needs to be changed is the fundamental ontology of the world. What you have to do is give up the idea that the material world consists of particles (since it turns out that a relativistic quantum theory of particles, a theory which satisfies (A)–(D), just can't be cooked up) and adopt the idea that it consists of something else. Here's the general idea:

What goes on in relativistic quantum theories is that one imagines that there is an infinitely tiny physical system permanently located at every single mathematical point in the entirety of space;

one imagines (that is) that there is literally an infinite array of such systems, one for each such point. And each one of those infinitely tiny systems is stipulated to be a quantum-mechanical system. And each one of them is stipulated to interact in a particular way with each of its neighbors.[13] And the complete array of them is called the field.

And it turns out that a relativistic quantum theory of the field (a theory which satisfies (A)–(D)) *can* be cooked up; moreover, that theory can be cooked up in such a way as to guarantee that the familiar quantum-mechanical observables of the material world (that is: the observables of "particles") can for the most part be reinterpreted there as observables of the field. For example, statements about the number of particles in a given region of space, or about the kinds of particles in that region, or about the physical states of the particles in that region, will all turn into statements about the quantum states of the infinitely tiny field systems in that region. And as a matter of fact, it can be shown that what the nonrelativistic limit of this theory amounts to is precisely the non-relativistic quantum theory of particles which was outlined in the section above on coordinate space and which most of this book is going to be about.

13. Or you could put it this way: What goes on in relativistic quantum theories is that one imagines that every single mathematical point in space is *itself* a quantum-mechanical system; and that each one of them interacts in a particular way with its neighbors.

··· 3 ···

Nonlocality

A famous attempt to escape from the standard way of thinking about quantum mechanics was initiated in the 1930s by Einstein and Podolsky and Rosen, and had a surprising aftermath, in the sixties, in the work of Bell, and that (Bell's work) is what this chapter will mostly be about.

But first let me describe the escape attempt itself. In 1935, Einstein, Podolsky, and Rosen (who have since then come to be known as "EPR") produced an argument, which was supposed to open the way to that escape, that (if the predictions of quantum mechanics about the outcomes of experiments are correct) the quantum-mechanical description of the world is necessarily *incomplete*.

Here's what they meant by "completeness": a description of the world is complete, for them, just in case nothing that's true about the world, nothing that's an "element of the reality" of the world, gets left out of that description.

Of course, that entails that if we want to find out whether or not a certain description of the world is complete, we need first to find out what all the elements of the reality of the world *are*; and it turns out (not surprisingly!) that EPR had nothing whatever to offer in the way of a general prescription for *doing* that. What they did (which is something much narrower, but which turns out to be enough for their purposes) is to write down a merely *sufficient* condition for a measurable property of a certain system at a certain moment to be an element of the reality of that system at that moment. The condition is that "if, without in any way disturbing a system, we can predict with certainty (i.e., with probability equal to unity) the value of a physical quantity, then there exists an element of reality corresponding to this physical quantity."

Let's see what this condition amounts to. Consider a question like this: If a measurement of a certain particular observable (call it O) of a certain particular physical system (call it S) were to be carried out at a certain particular future time (call it T), what would the outcome be? Suppose that there is a method whereby I can put myself in position, prior to T, to answer that question, with certainty. And suppose that the method whereby I can put myself in that position involves no physical disturbance of S whatsoever. Then (according to EPR) there must *now* already *be* some matter of fact about what the outcome of a future O measurement on S would be; there must *now* already *be* some fact about S (since the facts about S aren't going to get tampered with from the outside in the course of my putting myself in a position to answer the question about the O measurement) in virtue of which that future measurement would come out in that particular way.[1]

So, what EPR want to argue (once again) is that if the empirical predictions of quantum mechanics are correct, then there must be elements of the reality of the world which have no corresponding elements in the quantum-mechanical *description* of the world. They

---

1. Here are two particularly trivial examples. Suppose I have just now measured the color of some particular electron. Having done that measurement (since measurements of color are repeatable) puts me in a position to predict, with certainty, what the outcome of a measurement of the color of that electron at a later time would be, if such a measurement were to be carried out; and of course making such a prediction need not involve any further interaction with the electron at all. So the EPR reality condition entails that color must at present be an element of the reality of this electron; and of course that's also precisely what the quantum-mechanical formalism and the standard way of thinking entails.

Suppose, on the other hand, that I have just now measured the hardness of an electron. In order to be able to predict with certainty what the outcome of a future measurement of the color of *that* electron would be (if such a measurement were to be carried out), I would need to measure the color of that electron (I would need, that is, to interact with it, potentially to disturb it); and so the EPR reality condition *doesn't* entail that color is an element of the reality of *this* electron at present; and *that's* in accordance with the quantum-mechanical formalism and the standard way of thinking too.

want to use quantum mechanics, sort of paradoxically, against itself.

The argument goes something like this:

Consider a system consisting of two electrons. Electron 1 is located at position 1, and electron 2 is located at position 2. The spin-space state of these two electrons is the following (nonseparable) one:

$$(3.1) \qquad |A\rangle = \tfrac{1}{\sqrt{2}}|\text{black}\rangle_1|\text{white}\rangle_2 - \tfrac{1}{\sqrt{2}}|\text{white}\rangle_1|\text{black}\rangle_2$$

$|A\rangle$, like any quantum state, is necessarily an eigenstate of some complete observable of this pair of electrons. Call that observable O_A, and suppose that $O_A|A\rangle = |A\rangle$. Now, $|A\rangle$ has been written down in equation (3.1) in the color basis, but it happens to be an extraordinary mathematical fact about this particular state that (as the reader can explicitly verify for herself, by means of equation (2.21)) this state has precisely the same form if it's written down in the *hardness* basis. That is:

$$(3.2) \qquad |A\rangle = \tfrac{1}{\sqrt{2}}|\text{hard}\rangle_1|\text{soft}\rangle_2 - \tfrac{1}{\sqrt{2}}|\text{soft}\rangle_1|\text{hard}\rangle_2$$

And as a matter of fact it turns out that $|A\rangle$ retains precisely the same mathematical form if color (in equation (3.1)) or hardness (in equation (3.2)) is replaced by gleb or by scrad (which were defined in Chapter 2) or by any one of the continuous infinity of electron spin-space observables whatsoever.

Focus first on equation (3.1). Suppose that we were to carry out a measurement of the color of electron 1. The outcome of that measurement will be either "black" or "white," with equal probability (that follows from the rules for calculating probabilities for two-particle systems).[2] Moreover, quantum mechanics entails (and it is experimentally known to be true) that in the event that the outcome of that measurement is "black," then the outcome of any subsequent measurement of the color of *electron 2* will necessarily

2. See equation (2.38).

be "white," and in the event that the outcome of that measurement on particle 1 is "white," then the outcome of any subsequent measurement of particle 2 will necessarily be "black" (all of *that* follows from the collapse postulate for two-particle systems).[3]

EPR assumed (and this is the *only* assumption that enters into their argument other than the assumption that the predictions of quantum mechanics about the results of experiments are correct; and the name of this other assumption is *locality*) that things could in principle be set up in such a way as to guarantee that the measurement of the color of electron 1 produces no physical disturbance whatsoever in electron 2. That seemed almost self-evident. There seemed to be any number of ways to do it: you could, for example, separate the two electrons by some immense distance (since quantum mechanics predicts that none of what's been said here depends on how far apart the two electrons happen to be), or you could insert an impenetrable wall between them or build impenetrable shields around them (since quantum mechanics predicts that none of what's been said depends on what happens to lie between or around those two electrons), or you could set up any array of detectors you like in order to verify that no measurable signals ever pass from one of the electrons to the other in the course of the experiment (since quantum mechanics predicts that no such array, in such circumstances, whatever sorts of signals it may be designed to detect, will ever register anything).[4]

So, returning to (3.1): whenever $|A\rangle$ obtains, there is a means of predicting, with certainty, and (in principle, if locality is true, and if you set things up right) without disturbing electron 2, what the outcome of any subsequent measurement of the color of electron

---

3. See, in particular, equations (2.45) and (2.46).

4. Let's say a little more about precisely what the locality assumption amounts to: the assumption says that I can't punch you in the nose unless my fist gets to where your nose is.

Of course, something I do with my fist *far* from where your nose is can *cause* some *other* fist which is *near* where your nose is to punch you in the nose (something I do with my fist might be a signal to somebody else to punch you in the nose, for example); the assumption is just that if my fist never gets anywhere near your nose then I can't punch you in the nose *directly*, then it can't be *my* fist that punches you in the nose. And if something I do with my fist far from where your nose is *is* the cause of your getting punched in the nose, then (on this assumption) there must

2 will be. The way to do that is to measure the color of electron 1; since it's known that the outcome of any measurement of the color of electron 2 will invariably be the *opposite* of the outcome of any measurement of the color of electron 1. And so it follows from the reality condition that color must necessarily be an element of the reality of electron 2, that there must necessarily be a matter of fact about what the value of the color of electron 2 *is*, whenever $|A\rangle$ obtains.

Now focus on equation (3.2). Repeat precisely the same argument, with color replaced by hardness. It follows that whenever $|A\rangle$ obtains, there is a means of predicting, with certainty, and without disturbing electron 2, what the outcome of any subsequent measurement of the hardness of electron 2 will be (the way to do *that* is to measure the *hardness* of electron 1). And so it follows from the reality condition that the hardness of electron 2 is *also* necessarily an element of the reality of that particle whenever $|A\rangle$ obtains.

And so the standard way of thinking must simply be false, since there *can* be circumstances (like $|A\rangle$) in which there are simultaneously matters of fact about the values of both the color and the hardness of a single electron, even though those two observables are supposed to be incompatible.

---

necessarily be some causal sequence of events at contiguous points in space and at contiguous moments in time (the propagation of a signal, say) stretching all the way without a break from whatever it was that I did with my fist to your being punched in the nose. And of course the capacity of any such sequence of events to occur (the capacity, that is, of my fist to cause you to be punched in the nose) will necessarily depend on what sorts of physical conditions obtain in the space between my fist and your nose (it may, for example, depend on the absence of opaque walls, or of radio-wave absorbers, or what have you). And that sequence of events (in virtue of *being* a sequence, in virtue of being a string of causes whose effects are new causes whose effects are new causes . . .) must necessarily require some finite time to completely unfold.

And as a matter of fact the special theory of relativity entails that the velocities at which physical influences can be propagated through space by means of such sequences of neighboring events cannot possibly exceed the velocity of light (because otherwise there couldn't be any matter of objective fact about *what* the sequence of those events *is*); and that, of course, entails that there is yet another method (which the reader can now easily concoct for herself) of setting things up in such a way as to guarantee that the measurement of the color of electron 1 has no effect whatever on the outcome of the measurement of the color of electron 2.

Moreover, the formalism itself (quite apart from any particular way of *thinking* about it) must necessarily be *incomplete*, since there are necessarily certain facts, certain elements of the physical reality of the world, which have no corresponding elements in the formalism. There are facts about the color and the hardness of electron 2, for example, when $|A\rangle$ obtains, but there isn't anything in the mathematical description of the state $|A\rangle$, in this formalism, from which the values of the color or the hardness of electron 2 can be read off.

And since $|A\rangle$ retains its mathematical form in every basis, the same argument can be rehearsed in all of them, and so, when $|A\rangle$ obtains, there must necessarily simultaneously be matters of fact about the values of every spin-space observable of electron 2; and so the standard way of thinking must be (as it were) *infinitely* false, and the formalism must be infinitely incomplete.

And of course precisely the same arguments can be made about electron 1.

If all this is right, then whenever $|A\rangle$ obtains, all of the continuous infinity of spin observables are simultaneously elements of the realities of both of the electrons in question. And if that's so, then the statement that $|A\rangle$ obtains necessarily constitutes a very incomplete description of the state of a pair of electrons. The statement that $|A\rangle$ obtains must be true of a gigantic collection of *different* "true" states of the pair, in some of which, say, electron 1 is black and soft and scrad $= -1$ (and so on) and electron 2 is white and hard and scrad $= +1$ (and so on), and in others of which electron 1 is black and soft and scrad $= +1$ (and so on) and electron 2 is white and hard and scrad $= -1$ (and so on), and so on.

Nonetheless, the information that $|A\rangle$ obtains must certainly *constrain* the "true" state of a pair of electrons in a number of ways, since the outcomes of spin measurements on such pairs of electrons are (after all) determined by what their "true" states are, and since we're assuming that the quantum-mechanical predictions about the statistics of the outcomes of such measurements are correct.

Let's see what sorts of constraints arise. First of all, if $|A\rangle$ obtains (if, that is, the outcome of a measurement of O_A is $+1$), then the outcome of a measurement of any spin-space observable of electron 1 will necessarily be the opposite of the outcome of any measurement of the same observable on electron 2. Whenever $|A\rangle$ obtains, then, the "true" state of the pair of electrons in question is constrained, with certainty, to be one in which the *value* of every spin-space observable of electron 1 is the opposite of the value of that same observable of electron 2.

There are statistical sorts of constraints, too. Those are a bit more complicated: Suppose that $|A\rangle$ obtains, and suppose, for example, that we were to measure the color of electron 1 and then the scrad of electron 2. There will be four possible outcomes: black (color = +1) and scrad = − 1, white (color = −1) and scrad = +1, black and scrad = +1, and white and scrad = −1. Consider the first two, in both of which the value of the color of electron 1 is the opposite of the value of the scrad of electron 2. Let's calculate (in the way that we're instructed to by the quantum-mechanical formalism) the probability that the outcome of such an experiment would be either one of those. Initially $|A\rangle$ obtains. Then a measurement of the color of electron 1 is carried out. That measurement (in accordance with the probability rules and the collapse postulate for two-particle systems, which were both described in Chapter 2) will change the quantum state of this pair of electrons to either $|black\rangle_1|white\rangle_2$ or $|white\rangle_1|black\rangle_2$, with equal probabilities. In the first case (in the event, that is, that the outcome of the color measurement on electron 1 is "black"), the probability that the outcome of the scrad measurement on electron 2 is −1 is $|\langle white|scrad = -1\rangle|^2 = 1/4$. In the second case (in the event, that is, that the outcome of the color measurement on electron 1 is "white"), the probability that the outcome of the scrad measurement on electron 2 is +1 is $|\langle black|scrad = +1\rangle|^2 = 1/4$. Since the two possible outcomes of the color measurement of electron 1 are equally probable, it follows that the overall probability that the outcome of the color measurement of electron 1 is the opposite of the outcome of the scrad measurement of electron 2, when $|A\rangle$ obtains, is $1/4$.

And as a matter of fact, it turns out that if $|A\rangle$ obtains, and if either color or scrad or gleb are measured on both electrons, and if the observable that gets measured on electron 1 isn't the same observable as the one that gets measured on electron 2, then the quantum-mechanical probability that the outcome of the measurement on electron 1 is the opposite of the outcome of the measurement on electron 2 is always exactly ¼.

And that will amount to a constraint on the relative frequencies of various different "true" states of pairs of $|A\rangle$-type electrons, which runs as follows: Consider a large collection of pairs of electrons, each of which (each pair, that is) is known to have the property that $O_A = +1$. Pick any two of the three observables color, scrad, gleb. Call one P and the other Q. One-quarter (statistically) of the pairs of electrons in any such large collection will have to have the property that the value of P for electron 1 is the opposite of the value of Q for electron 2, and the remaining three-quarters of the pairs will have to have the property that the value of P for electron 1 is the same as the value of Q for electron 2.

And now (here comes the punch line) a well-defined question can be posed as to whether these two constraints (the deterministic constraint about the values of identical observables for the two electrons, and the statistical constraint about the values of different observables for the two electrons) are *mathematically consistent* with one another. It was Bell who first clearly posed and answered that question, twenty-nine years after the publication of the EPR argument (Bell, 1964); and it turns out that the answer to that question is no.[5]

And so the conclusion of the EPR argument is logically impossible; and so either locality must be false or the predictions of quan-

5. Here's why: Consider (to begin with) a collection of pairs of electrons which satisfies the deterministic constraint. In each of the pairs in any collection like that, the color value of electron 1 has to be the opposite of the color value of electron 2, and the scrad value of electron 1 has to be the opposite of the scrad value of electron 2, and the gleb value of electron 1 has to be the opposite of the gleb value of electron 2. Now, it turns out (as the reader can easily verify for herself) that there are exactly eight different ways in which values of color and gleb and scrad can possibly be assigned to a pair of electrons such that everything in the last sentence is true.

An explicit inspection of those eight possible assignments (which the reader can also easily accomplish on her own) will show that every single one of them (which is to say: every single one of the pairs of electrons in any collection of pairs of

tum mechanics about the outcomes of spin measurements on $|A\rangle$-states must be false (since those are the only two assumptions on which the argument depends); and it happens that those predictions are now experimentally known to be true; and so the assumption that the physical workings of the world are invariably local must (astonishingly) be false.

Here's another way to tell the story:

Einstein, Podolsky, and Rosen noticed that there was something odd about the collapse postulate for two-particle systems. They noticed that it was nonlocal: if the two particles are initially in a nonseparable state, then a measurement carried out on one of them can bring about changes, instantaneously, in the quantum-mechanical description of the *other* one, no matter how far apart those two particles may happen to be or what might lie between them.

Consider, for example, a pair of electrons, and suppose that $|A\rangle$ initially obtains and that a measurement of the color of electron 1 is carried out. The outcome of that measurement (as we've seen) will be either "black" or "white," with equal probabilities. The collapse postulate for two-particle systems entails that as of the instant that that measurement is over, the state of electron 2 will be either $|\text{white}\rangle$ (in the event that the outcome of the measurement on electron 1 was "black") or $|\text{black}\rangle$ (in the event that the outcome of the measurement on electron 1 was "white").

---

electrons whatever which satisfies the deterministic constraint) either has the property that the color value of electron 1 is the opposite of the gleb value of electron 2, or else it has the property that the color value of electron one is the opposite of the scrad value of electron 2, or else it has the property that the gleb value of electron 1 is the opposite of the scrad value of electron 2.

But note that in any collection of pairs of electrons which satisfies the *statistical* constraint, the fraction of the pairs which have the first one of the properties described in the last paragraph has got to be ¼, and the fraction of the pairs which have the second one of those properties has also got to be ¼, and the fraction of the pairs which have the third one of those properties has also got to be ¼; and so the fraction of the pairs of electrons which has *any* one of those three properties has got to be less than or equal to ¾.

And so *no* collection of pairs of electrons whatever can possibly satisfy both the statistical constraint and the deterministic one. Period.

EPR suspected that this nonlocality must merely be a disposable artifact of this particular mathematical formalism, of this particular procedure for calculating the statistics of the outcomes of experiments; and that there must be other (as yet undiscovered) such procedures, which give rise to precisely the same statistical predictions but which are entirely *local*.

And it emerged thirty years later, in the work of Bell, that that suspicion was demonstrably wrong.

Bell's work is sometimes taken to amount to a proof that any attempt to escape from the standard way of thinking, any attempt to be realistic about the values of the spin observables of a pair of electrons for which $|A\rangle$ obtains, must necessarily turn out to be nonlocal. But (and this is the point of telling the story this way) things are actually a good deal more serious than that. What Bell has given us is a proof that there is as a matter of fact a genuine nonlocality in the actual workings of nature, *however* we attempt to describe it, period. That nonlocality is, to begin with, a feature of quantum mechanics itself, and it turns out (via Bell's theorem) that it is necessarily also a feature of every possible manner of calculating (without or with superpositions) which produces the same statistical predictions as quantum mechanics does; and those predictions are now experimentally known to be correct.

Let's be somewhat more precise about just what sort of nonlocality quantum mechanics exhibits.

First of all, when $|A\rangle$ obtains, the statistics of the outcomes of spin measurements on electron 2 depend nonlocally (as we've seen) on the outcomes of spin measurements on electron 1, and vice versa. But consider whether or not the statistics of the outcomes of spin measurements on electron 2, when $|A\rangle$ obtains, depend nonlocally on *whether* a spin measurement is first carried out on electron 1 (and vice versa).

Let's figure it out. Suppose that $|A\rangle$ obtains, and suppose, to begin with, that a measurement of color is carried out on electron 2. Well, it follows from equation (3.1), and from the standard quantum-mechanical rules for calculating the probabilities of measurement outcomes, that the outcome of that measurement is equally likely to be "black" or "white." And now suppose that $|A\rangle$ obtains and that a measurement of the color of electron 1 is carried out, and *then* a measurement of the color of electron 2 is carried out. Well, the measurement of the color of electron 1 is now (for precisely the same reasons) equally likely to come out "black" or "white," In the event that it comes out "black" it follows from the collapse postulate for two-particle systems that the subsequent measurement of the color of electron 2 will come out "white," and in the event that it comes out "white" it follows from the collapse postulate for two-particle systems that the subsequent measurement of electron 2 will come out "black." So, when $|A\rangle$ obtains, the outcome of a measurement of the color of electron 2 is equally likely to be "black" or "white" *whether or not* a measurement of the color of electron 1 is carried out first.

Now suppose that $|A\rangle$ obtains and that a measurement of the hardness of electron 1 is carried out, and then a measurement of the color of electron 2 is carried out. It follows from equation (3.2), and from the probability rules, that the outcome of the hardness measurement on electron 1 is equally likely to be "hard" or "soft." Now, in the event that the outcome of that first measurement is "soft," it follows from the collapse postulate and from the probability rules that the outcome of the second measurement (the color measurement on electron 2) is equally likely to be "black" or "white." And the same thing is true in the event that the outcome of the first measurement is "hard." And so here's where we are so far: when $|A\rangle$ obtains, the outcome of a measurement of the color of electron 2 is equally likely to be "black" or "white," whether a measurement of the color of electron 1 is carried out first, or a measurement of the hardness of electron 1 is carried out first, or no measurement on electron 1 is carried out first.

And everything that's been said here is also true if "color" and "hardness" are interchanged, or if either or both of them are replaced by "gleb" or by "scrad" or by any of the other spin observables; and everything that's been said here is of course also true if "electron 1" is exchanged with "electron 2." You can extend this kind of argument to other sorts of observables and to other sorts of physical systems, too.

And if you follow this sort of thing out as far as you can, here's where you end up: Take any composite physical system S. Divide it up, any way you like, into two (possibly still composite) subsystems. Call those two subsystems s_1 and s_2. Take any quantum state whatsoever, $|Q\rangle$, of S. Choose any observable O_1 of s_1 and any observable O_2 of s_2. It's possible to prove the following completely general theorem: Whenever $|Q\rangle$ obtains, the probabilities of the various possible outcomes of a measurement of O_2 don't depend at all on whether or not a measurement of O_1 is carried out first.

So, there are (for the nth time) nonlocal influences in nature (if the relevant predictions of quantum mechanics are right; and they are right); but those influences are invariably of a particularly subtle kind. The outcomes of measurements do sometimes depend nonlocally on the outcomes of other, distant, measurements; but the outcomes of measurements invariably *do not* depend nonlocally on *whether* any other, distant, measurements get carried out!

Let me put it another way (and this is really the punch line of this little section). The subtlety of these influences is such that (even though they surely exist, even though the statistics of the outcomes of experiments can't be understood without them) *they cannot possibly be exploited to transmit a detectable signal, they cannot possibly be made to carry information*, nonlocally, between any two distant points. The problem is that it just won't work to encode the information you want to transmit in a decision to make a measurement or not to make one, or in a decision about what to make a measurement of, since (as we've just seen) no such decisions can ever have detectable nonlocal effects; and there just isn't any *way* to encode the information you want to transmit in the *outcome* of an appropriate sort of experiment (in, say, the outcome of a measurement of the color of electron 1, when $|A\rangle$ obtains), since, as a matter of principle, the outcomes of such measurements are necessarily entirely beyond our control; and all of the rest of quantum mechanics, the parts that deal with the reactions of physical systems to everything other than measurements (that is: the dynamical equations of motion), are, so far as we know, absolutely local from start to finish.

There will be a good deal more to say of all this later on.

··· 4 ···

The Measurement Problem

Now I want to begin to worry in earnest about whether or not the dynamics says the same thing as the postulate of collapse says about what happens to the state vector of a physical system when the system gets measured. Here's what looked worrisome back in Chapter 2: the dynamics (which is supposed to be about how the state vectors of physical systems evolve *in general*) is fully deterministic, but the collapse postulate (which is supposed to be about how the state vector of a system evolves when it comes in contact with a measuring device) isn't; and so it isn't clear precisely how the two can be consistent.

Let's figure out what the dynamics says about what happens when things get measured.

Suppose that everything in the world always evolves in accordance with the dynamical equations of motion. And suppose that we have a device (which operates in accordance with those equations, just like everything else does) for measuring the hardness of an electron; and suppose that that device works like this: The device has a dial on the front, with a pointer; and the pointer has three possible positions. In the first position the pointer points to the word "ready," and in the second position it points to the word "hard," and in the third it points to the word "soft." Electrons are fed into one side of the device and come out the other, and in the course of passing through (if the device is set up right, with the pointer initially in its "ready" position) they get their hardnesses measured, and the outcomes of those measurements get recorded in the final position of the pointer (figure 4.1).

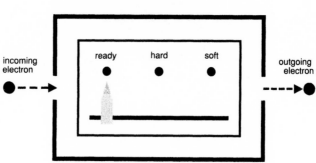

Figure 4.1

If the device is set up right, and if the dynamics is always true, then (to put all this another way) the dynamical equations of motion entail that it behaves like this:

$$(4.1) \qquad |\text{ready}\rangle_m|\text{hard}\rangle_e \rightarrow |\text{``hard''}\rangle_m|\text{hard}\rangle_e$$

and

$$(4.2) \qquad |\text{ready}\rangle_m|\text{soft}\rangle_e \rightarrow |\text{``soft''}\rangle_m|\text{soft}\rangle_e$$

That is: if the device (whose state vector is labeled with subscript m) is initially in the ready state, and if an electron (whose state vector is labeled with subscript e) that is hard gets fed through it, then the device ends up in the state wherein the pointer is pointing at "hard"; and if the device is initially in the ready state, and if a *soft* electron gets fed through it, then the device ends up in the state wherein the pointer is pointing at "soft." That's what it *means* for a measuring device for hardness to be a good one and to be set up right.

Now (still supposing that the dynamics is always true), consider what happens if this device (the one for measuring *hardness*) is set up right, and is in its ready state, and a *black* electron is fed into it. It turns out that (4.1) and (4.2), and the fact that the dynamical equations of motion are invariably linear, suffice by themselves to figure that out: The initial state of the electron and the measuring device is

(4.3) $|\text{ready}\rangle_m|\text{black}\rangle_e = |\text{ready}\rangle_m\{1/\sqrt{2}|\text{hard}\rangle_e + 1/\sqrt{2}|\text{soft}\rangle_e\}$

$= 1/\sqrt{2}|\text{ready}\rangle_m|\text{hard}\rangle_e + 1/\sqrt{2}|\text{ready}\rangle_m|\text{soft}\rangle_e$

which is precisely $1/\sqrt{2}$ times the initial state in (4.1) plus $1/\sqrt{2}$ times the initial state in (4.2). So, since (by hypothesis) the dynamical equations of motion entail that $|\text{ready}\rangle_m|\text{hard}\rangle_e$ evolves as in (4.1) and that $|\text{ready}\rangle_m|\text{soft}\rangle_e$ evolves as in (4.2) (that is: since this device is set up right, and since it's a good measuring device for hardness), it follows from the linearity of those equations that the state in (4.3), when the measuring device gets switched on, will necessarily evolve into

(4.4) $1/\sqrt{2}|\text{"hard"}\rangle_m|\text{hard}\rangle_e + 1/\sqrt{2}|\text{"soft"}\rangle_m|\text{soft}\rangle_e$

That's how things end up, with certainty, according to the dynamics.[1]

And the way things end up according to the postulate of *collapse* (when you start with (4.3)) is

(4.5) *either* $|\text{"hard"}\rangle_m|\text{hard}\rangle_e$ (with probability $1/2$)

or $|\text{"soft"}\rangle_m|\text{soft}\rangle_e$ (with probability $1/2$)

1. This result often strikes people as mysteriously easy. The intuition is that measuring devices for hardness like the one described here must be extremely complicated contraptions (especially if you look at them on the level, say, of their constituent atoms) and must have extremely complicated equations of motion, the solution of which must be an extremely complicated matter. All of that is true. What simplifies things here is the fact that however complicated those equations may be, (4.1) and (4.2) are surely solutions of them (since the contraption we're dealing with is, by hypothesis, and whatever *else* it may be, a good measuring device for the hardness of an electron), and they (the equations) are surely *linear;* and those two facts are enough by themselves to insure that if a black electron is fed into the device, then those equations will entail that things will end up in the state in (4.4).

And the trouble is that (4.4) and (4.5) are measurably different situations.[2]

The state described in (4.5) is the one that's right; it is (as a matter of empirical fact) how things *do* end up when you start with (4.3).

The state described in (4.4) is *not* how things end up;[3] (4.4) is something very strange. It's a superposition of one state in which the pointer is pointing at "hard" and another state in which the pointer is pointing at "soft"; it's a state in which (on the standard way of thinking) *there is no matter of fact* about where the pointer is pointing.[4]

Let's make this somewhat sharper. Suppose that a human observer enters the picture, and *looks* at the measuring instrument (when the measurement is all done) and *sees* where the pointer is pointing. Let's figure out what the dynamics will say about that.

Suppose, then (just as we did before), that literally every physical system in the world (and this now includes human beings; and it

2. Perhaps this ought to be expanded on a bit. The point is that there are (in accordance with the postulates of quantum mechanics that were laid out in Chapter 2) necessarily measurable properties of the state in (4.4) whereby it can, in principle, be experimentally distinguished from either of the states in (4.5) and, as a matter of fact, from any other state whatever. There will be a good deal to say, later on, about precisely *what* those properties *are* (they're complicated ones, and their measurement will in general be extremely difficult); what's important for the moment is simply that those properties exist.

3. It isn't, at any rate, according to the conventional wisdom about these matters; but there will be much more to say about this later on.

4. Something ought to be mentioned in passing here, something that will turn out to be important later on.

What we've just discovered is that there is a certain fundamental effect of the carrying out of a measurement (namely: the emergence of some definite *outcome* of the measurement; the emergence of some *matter of fact* about precisely *what* the outcome of the measurement *is*) which is not predicted by the dynamical equations of motion.

But consider another effect of the carrying out of a measurement, one which we first described in the course of our discussions of hardness and color in Chapter 1: The carrying out of a measurement is disruptive of the values of observables of the measured system which are incompatible with the observable that gets measured. It turns out that the dynamical equations of motion *do* predict *that*.

includes the brains of human beings) always evolves in accordance with the dynamical equations of motion; and suppose that a black electron is fed through a measuring device for hardness that's set up right and that starts out in its ready state (so that the state of the electron and the device is now the one in (4.4)); and suppose that somebody named Martha comes along and looks at the device; and suppose that Martha is a competent observer of the positions of pointers.

Being a "competent observer" is something like being a measuring device that's set up right: What it means for Martha to be a competent observer of the position of a pointer is that whenever Martha looks at a pointer that's pointing to "hard," she eventually comes to *believe* that the pointer is pointing to "hard"; and that whenever Martha looks at a pointer that's pointing to "soft," she eventually comes to believe that the pointer is pointing to "soft" (and so on, in whatever direction the pointer may be pointing). What it means (to put it somewhat more precisely) is that the dynamical equations of motion entail that Martha (who is a physical system, subject to the physical laws) behaves like this:

(4.6) $|ready\rangle_o|ready\rangle_m \rightarrow |\text{"ready"}\rangle_o|ready\rangle_m$

and

          $|ready\rangle_o|\text{"hard"}\rangle_m \rightarrow |\text{"hard"}\rangle_o|\text{"hard"}\rangle_m$

and

          $|ready\rangle_o|\text{"soft"}\rangle_m \rightarrow |\text{"soft"}\rangle_o|\text{"soft"}\rangle_m$

In these expressions, $|ready\rangle_o$ is that physical state of Martha's brain in which she is alert and in which she is intent on looking at the

---

Look, for example, at the evolution from (4.3) to (4.4). Equation (4.3) is an eigenstate of the *color* of the electron whose hardness is about to be measured; but (4.4) (which is the state following the interaction of the electron with a good measuring device for the hardness, according to the equations of motion) *isn't*. If (4.4) is written out in terms of eigenstates of the color of the electron (which the reader can now easily do), *it* turns out to be a superposition of two states (with equal coefficients), in one of which the electron is black and in the *other* of which the electron is white.

pointer and finding out where it's pointing; $|\text{"ready"}\rangle_o$ is that phys-
ical state of Martha's brain in which she believes that the pointer
is pointing to the word "ready" on the dial; $|\text{"hard"}\rangle_o$ is that
physical state of Martha's brain in which she believes that the
pointer is pointing to the word "hard" on the dial; and $|\text{"soft"}\rangle_o$ is
that physical state of Martha's brain in which she believes that the
pointer is pointing to the word "soft" on the dial.[5]

Let's get back to the story. The state of the electron and the
measuring device (at the point where we left off) is the strange one
in (4.4). And now in comes Martha, and Martha is a competent
observer of the position of the pointer, and Martha is in her ready
state, and Martha looks at the device. It follows from the linearity
of the dynamical equations of motion (if those equations are right),
and from what it means to be a competent observer of the position
of the pointer, that the state when Martha's done is with certainty
going to be

$$(4.7) \qquad 1/\sqrt{2}|\text{"hard"}\rangle_o|\text{"hard"}\rangle_m|\text{hard}\rangle_e + 1/\sqrt{2}|\text{"soft"}\rangle_o|\text{"soft"}\rangle_m|\text{soft}\rangle_e$$

That's what the dynamics entails.

And of course what the postulate of *collapse* entails is that when
Martha's all done, then

$$(4.8) \qquad either \quad |\text{"hard"}\rangle_o|\text{"hard"}\rangle_m|\text{hard}\rangle_e \quad \text{(with probability } 1/2)$$
$$or \quad |\text{"soft"}\rangle_o|\text{"soft"}\rangle_m|\text{soft}\rangle_e \quad \text{(with probability } 1/2)$$

is going to obtain.

And (4.7) and (4.8) are empirically different. The state described
in (4.8) is the one that's right; (4.7) is unspeakably strange. The
state described in (4.7) is at odds with what we know of ourselves

5. It hardly needs saying that this is an absurdly oversimplified description of
Martha's brain, and that this is an absurdly oversimplified account of the ways in
which mental states are generally supposed to supervene on brain states; but all
that turns out not to make any difference (not at *this* stage of the game, anyway).
We can fill in the details whenever we want, to whatever extent we want. They
won't change the arguments.

by *direct introspection*. It's a superposition of one state in which Martha thinks that the pointer is pointing to "hard" and another state in which Martha thinks that the pointer is pointing to "soft"; *it's a state in which there is no matter of fact about whether or not Martha thinks the pointer is pointing in any particular direction.*[6]

And so things are turning out badly. The dynamics and the postulate of collapse are flatly in contradiction with one another (just as we had feared they might be); and the postulate of collapse seems to be right about what happens when we make measurements, and the dynamics seems to be bizarrely *wrong* about what happens when we make measurements; and yet the dynamics seems to be *right* about what happens whenever we *aren't* making measurements; and so the whole thing is very confusing; and the problem of what to do about all this has come to be called "the problem of measurement."

We shall be thinking about that for the rest of this book.

6. This isn't anything like a state in which Martha is, say, *confused* about where the pointer is pointing. *This* (it deserves to be repeated) is something *really strange*. This is a state wherein (in the language we used in Chapter 1) it isn't right to say that Martha believes that the pointer is pointing to "hard," and it isn't right to say that Martha believes that the pointer is pointing to "soft," and it isn't right to say that she has *both* of those beliefs (whatever *that* might mean), and it isn't right to say that she has neither of those beliefs.

··· 5 ···

The Collapse of the Wave Function

The Idea of the Collapse

The measurement problem was first put in its sharpest possible form in the 1930s, by John von Neumann, in an extraordinary book called *Mathematical Foundations of Quantum Mechanics* (von Neumann, 1955). It looked to von Neumann as though the only thing that could possibly be done about the measurement problem was to bite the bullet, and admit that the dynamics is simply wrong about what happens when measurements occur, and nonetheless right about everything else. And so what he concluded was that there must be two fundamental laws about how the states of quantum-mechanical systems evolve:

I. When no measurements are going on, the states of all physical systems invariably evolve in accordance with the dynamical equations of motion.
II. When there *are* measurements going on, the states of the measured systems evolve in accordance with the postulate of collapse, *not* in accordance with the dynamical equations of motion.

But this clearly won't do. Here's the trouble: What these laws actually *amount to* (that is: what they actually *say*) will depend on the precise meaning of the word *measurement* (because these two laws entail that *which one* of them is being *obeyed* at any given moment depends on whether or not a "measurement" is being *carried out* at that moment). And it happens that the word *mea-*

surement simply doesn't have any absolutely precise meaning in ordinary language; and it happens (moreover) that von Neumann didn't make any attempt to cook up a meaning for it, either.

And so those laws, as von Neumann wrote them down, simply don't determine exactly how the world behaves (which is to say: they don't really amount to prospective fundamental "laws" at all).

And there has consequently been a long tradition of attempts to figure out how to write them down in such a way that they *do*.

Here's where things stand: Suppose that a certain system is initially in an eigenstate of observable A, and that a measurement of observable B is carried out on that system, and that A and B are incompatible with one another. What we know with absolute certainty, by pure introspection, is that by the time that measurement is all done, and a sentient observer has looked at the measuring device and formed a conscious impression of how that device presently appears and what it presently indicates, then some wave function must already have violated the dynamical equations of motion and collapsed. What we need to do is to figure out precisely when that collapse occurs.

Let's try to guess.

Perhaps the collapse always occurs precisely at the last possible moment; perhaps (that is) it always occurs precisely at the level of consciousness,[1] and perhaps, moreover, consciousness is always the agent that brings it about.

Put off the temptation to dismiss this as nonsense just for long enough to see what it amounts to.

On this proposal (which is due to Wigner, 1961), the correct laws of the evolution of the states of physical systems look something like this: All physical objects almost always evolve in strict accordance with the dynamical equations of motion. But every now and then, in the course of some such dynamical evolutions (in the course

---

1. This isn't a way of saying that there's anything *illusory* about the collapse; it's just a way of saying at precisely what point the collapse (which is a *physical* process) occurs, a way of saying precisely what sorts of processes precipitate collapses.

of *measurements,* for example), the brain of a sentient being may enter a state wherein (as we've seen) states connected with various different *conscious experiences* are superposed; and at such moments, the *mind* connected with that brain (as it were) *opens its inner eye,* and *gazes* on that brain, and that causes the entire system (brain, measuring instrument, measured system, everything) to collapse, with the usual quantum-mechanical probabilities, onto one or another of those states;[2] and then the eye closes, and everything proceeds again in accordance with the dynamical equations of motion until the next such superposition arises, and then that mind's eye opens up again, and so on.

This proposal entails that there are two fundamentally different sorts of physical systems in the world:

A. *Purely physical* systems (that is: systems which *don't* contain sentient observers). These systems, so long as they remain isolated from outside influences, always evolve in accordance with the dynamical equations of motion.
B. *Conscious* systems (that is: systems which *do* contain sentient observers). These systems evolve in accordance with the more complicated rules described above.[3]

But the trouble here is pretty obvious too: How the physical state of a certain system evolves (on this proposal) depends on whether or not that system is *conscious*; and so in order to know precisely how things physically behave, we need to know precisely what is

2. Here's an example: An observer carries out a measurement of the hardness of a black electron. Eventually (when the measuring device has done its work, and the observer has looked at the device), things get to be in the state in equation (4.7), and then the mind's eye of the observer opens, gazes upon her brain, and causes a collapse, with equal probabilities, onto either the first or the second of the terms in that state.

3. This (needless to say) amounts to a full-blown radically interactive mind-body dualism. Wigner thought that this sort of dualism turns out (ironically) to be a *necessary consequence of physics:* he thought that there was a physical job to be done in the world (the job of collapsing wave functions) which could only be done by a not-purely-physical thing.

conscious and what isn't. What this "theory" predicts (that is: what "theory" it *is*) will hinge on the precise meaning of the word *conscious;* and that word simply doesn't have any absolutely precise meaning in ordinary language; and Wigner didn't make any attempt to make up a meaning for it; and so all this doesn't end up amounting to a genuine physical theory either.

Let's try another guess. Perhaps the collapse occurs at the level of *macroscopicness* (not at the last possible moment but, as it were, at the last reasonable moment). On this proposal (which has lots of originators and lots of adherents), the correct laws of the evolution of the states of physical systems look something like this: All physical objects almost always evolve in strict accordance with the dynamical equations of motion. But every now and then, in the course of some such dynamical evolutions (in the course of measurements, for example), it comes to pass (as we've seen) that two macroscopically different conditions of a certain system (two different orientations of a pointer, say) get superposed, and at that point, as a matter of fundamental physical law, the state of the entire system collapses, with the usual quantum-mechanical probabilities, onto one or another of those macroscopically different states.[4] Then everything proceeds again in accordance with the dynamical equations of motion until the next such superposition arises, and then another such collapse takes place, and so on.

There are two sorts of physical systems in the world according to this proposal too:

A'. Purely *microscopic* systems (that is: systems which *don't* contain macroscopic subsystems). These systems, so long as they remain isolated from outside influences, always evolve in accordance with the dynamical equations of motion.

B'. *Macroscopic* systems (that is: systems which *do* contain macroscopic subsystems). These systems evolve in accordance with the more complicated rules described above.

---

4. Here's an example: Somebody carries out a measurement of the hardness of a black electron. Eventually (when the measuring device has done its work), things get to be in the state in equation (4.4), and then that state collapses, with equal probabilities, onto either the first or the second of the terms in that state.

The trouble *here* is going to be about the meaning of *macroscopic*. And that will put us back (for the third time now) where we started.[5] And so all this is getting us nowhere.

We need to find a less ambiguous way to talk.

A Digression on Why It's Hard to Settle the Question Empirically

Let's see if we can settle the question empirically.

Here's the idea: The collapse of the wave function (whatever else it might turn out to be) is, after all, a physical event, with physical consequences; and those consequences must in principle be detectable; and so the question of precisely where and when collapses occur (which is what we've been merely guessing about over the last few pages) must in principle be answerable, with certainty, by means of the right sorts of experiments.

Suppose (for example) that I have a theory about the collapse; and suppose that my theory entails that if I pass a black electron through a measuring device for hardness like the one described in Chapter 4, then at precisely the moment when the state of the electron and the measuring device becomes (in accordance with the dynamics) the one in equation (4.4), that state collapses, with the usual quantum-mechanical probabilities, onto one or the other of the two terms in that state. And suppose that my friend has another theory about the collapse; and suppose that *her* theory entails that the collapse doesn't happen until some particular *later* moment, some moment *farther on* in the measuring process (only when, say, a human retina gets involved, or an optic nerve, or a brain, or whatever). And suppose that we should like to test these two theories against one another by means of an experiment.

Here's how to start: Feed a black electron into a measuring device for hardness and give it enough time to pass all the way through. If *my* theory is right, then the state of the electron and the measur-

---

5. There is, as a matter of fact, an astonishingly long and bombastical tradition in theoretical physics of formulating these sorts of guesses about precisely when the collapse occurs in language which is so imprecise as to be (as we've just seen) absolutely useless. Some of the words that come up in these guesses (besides *measurement* and *consciousness* and *macroscopic*) are *irreversible, recording, information, meaning, subject, object,* and so on.

ing device at that moment (the moment at which the electron has had just enough time to pass all the way through the device) ought to be

(5.1)   *either* $|\text{"hard"}\rangle_m|\text{hard}\rangle_e$ (with probability ½)

         or $|\text{"soft"}\rangle_m|\text{soft}\rangle_e$ (with probability ½)

But if my *friend's* theory is right, *then* the state at that same moment ought to be

(5.2) $\frac{1}{\sqrt{2}}|\text{"hard"}\rangle_m|\text{hard}\rangle_e + \frac{1}{\sqrt{2}}|\text{"soft"}\rangle_m|\text{soft}\rangle_e$

just as the dynamics predicts (the *violation* of the dynamics, in my friend's theory, doesn't happen until later). And so now all we need to do is to figure out a way to distinguish, by means of a measurement, between the circumstances in (5.1) (wherein the pointer is pointing in some particular, but as yet unknown, direction) and the circumstances in (5.2) (wherein the pointer *isn't* pointing in any particular direction *at all*).

Let's figure out what sort of an observable we would need to measure in order to do that.

What if we were to measure the position of the tip of the pointer (that is: what if we were to measure *where* the pointer is *pointing*)?[6] Here's why that won't work: If my theory is right and, consequently, (5.1) obtains, then of course a measurement of the position of the tip of the pointer will have a fifty-fifty chance of finding the pointer in the "pointing-at-hard" state, and it will have a fifty-fifty chance of finding the pointer in the "pointing-at-soft" state (since, on my theory, the pointer is now already *in* one of those two states, each with probability ½). But if my *friend's* theory is right and, consequently, (5.2) obtains, *then* a measurement of the position of

6. And note that we shall be supposing in what follows that *this* measurement (the measurement of the position of the tip of the pointer on the hardness measuring device) gets carried through all the way to the end: all the way to the point where some observer becomes aware of the outcome of that measurement. By the time all that gets done, even my friend's theory about the collapse (that is: *any* theory about the collapse whatever) will have to entail that a collapse has already occurred.

the tip of the pointer will have a fifty-fifty chance of *collapsing* the state vector of the pointer onto the "pointing-at-hard" state, and a fifty-fifty chance of collapsing it onto the "pointing-at-soft" state.[7] And so the probability of any given outcome of a measurement of the position of the pointer will be the *same* on these two theories; and so this isn't the sort of measurement we're looking for.

What about measuring the hardness of the electron? That won't work either. If my theory is right and (5.1) obtains, then a measurement of the hardness of the electron will have a fifty-fifty chance of finding the electron in the hard state and a fifty-fifty chance of finding it in the soft state; and if my friend's theory is right and (5.2) obtains, then a measurement of the hardness of the electron will have a fifty-fifty chance of collapsing the state vector of the electron onto the hard state and a fifty-fifty chance of collapsing it onto the soft state. Again, the probabilities will be the same on both theories, and so this isn't the sort of measurement we're looking for, either.

What about measuring the *color* of the electron? That won't work either. On my theory, one of the two states in (5.1) now obtains and a measurement of the color of the electron in *either one* of those states will have a fifty-fifty chance of collapsing the state of the electron onto $|black\rangle_e$ and a fifty-fifty chance of collapsing the state of the electron onto $|white\rangle_e$. On my friend's theory, the state in (5.2) now obtains and a measurement of the color of the electron in *that* state will likewise have a fifty-fifty chance of collapsing the state of the electron onto $|black\rangle_e$ and a fifty-fifty chance of collapsing it onto $|white\rangle_e$.[8]

---

7. Of course, in the event that the state of the pointer gets collapsed onto the "pointing-at-hard" state (onto $|\text{"hard"}\rangle_m$, that is), then the state of the electron will automatically get collapsed onto $|hard\rangle_e$; and in the event that the state of the pointer gets collapsed onto the "pointing-at-soft" state ($|\text{"soft"}\rangle_m$), then the state of the electron gets collapsed onto $|soft\rangle_e$. All of that follows from (5.2), together with the collapse postulate for composite systems, which was spelled out in Chapter 2.

8. This is, as a matter of fact, something that we've noted before, in a slightly different language, in note 4 of Chapter 4. The reader can easily confirm it (as we mentioned there) by writing out the state in (5.2) in terms of eigenstates of the color of the electron.

Let's go back to the device. Consider an observable of the device which (for lack of any appropriate name) we can call zip. The eigenstates of zip are as follows:[9]

(5.3) $|zip = 0\rangle = |ready\rangle_m$

$|zip = +1\rangle = \frac{1}{\sqrt{2}}|\text{"hard"}\rangle_m + \frac{1}{\sqrt{2}}|\text{"soft"}\rangle_m$

$|zip = -1\rangle = \frac{1}{\sqrt{2}}|\text{"hard"}\rangle_m - \frac{1}{\sqrt{2}}|\text{"soft"}\rangle_m$

Measuring zip won't work either, since (much like with the color of the electron) if either of the states in (5.1) obtains, or if the state in (5.2) obtains, a measurement of zip has a fifty-fifty chance of collapsing the state of the pointer onto the state $|zip = +1\rangle$ and a fifty-fifty chance of collapsing the state of the pointer onto the $|zip = -1\rangle$ state.[10]

9. A few remarks are in order here.

First of all, it follows from property (5) of Hermitian operators (in Chapter 2) that some such observable as zip (that is: some such *Hermitian operator,* some operator with precisely these eigenstates) necessarily exists.

One of the eigenstates of zip (the zip = 0 state) is of course also an eigenstate of the pointer-position, but zip is nonetheless in general incompatible with the pointer-position: the zip = +1 state and the zip = −1 state are both superpositions of states in which the pointer is pointing in different directions.

The matrix for zip in the pointer-position basis

(that is: the basis in which $|ready\rangle_m = \begin{bmatrix} 1 \\ 0 \\ 0 \end{bmatrix}$, $|\text{"hard"}\rangle_m = \begin{bmatrix} 0 \\ 1 \\ 0 \end{bmatrix}$, and $|\text{"soft"}\rangle_m = \begin{bmatrix} 0 \\ 0 \\ 1 \end{bmatrix}$)

will be zip $= \begin{bmatrix} 1 & 0 & 0 \\ 0 & 0 & 1 \\ 0 & 1 & 0 \end{bmatrix}$

The reader may want to calculate the matrix for the pointer-position in this same basis (it's very easy), and to confirm that that latter matrix doesn't commute with the matrix for zip.

10. Note, by the way, that everything that's been discovered here about the state in (5.2) could have been surmised, if we had been cleverer, right at the outset. The state in (5.2) is, after all, a *nonseparable* state of the electron and the measuring device, a state in which no observable of the electron alone or of the measuring device alone can *possibly* have any particular value.

Consider, finally, an observable of the composite system consisting of the measuring device *and* the electron, zip − color (that is: zip *minus* color). It turns out that the state in (5.2) *is* an eigenstate of zip − color (its associated eigenvalue is 0),[11] and that *neither* of the states in (5.1) are eigenvalues of zip − color. And so, if my theory is right and (5.1) obtains, then a measurement of zip − color (or, rather, a collection of such measurements, carried out on similarly prepared systems) will have a statistical distribution of various *different* possible outcomes; but if my *friend's* theory is right and (5.2) obtains, then the outcome of every measurement of zip − color will necessarily, with certainty, be 0. And so my friend's theory and my theory *can* be distinguished from one another, *empirically*, by means of measurements of zip − color.

And this, needless to say, is a very general sort of fact: any claim about precisely where and precisely when collapses occur can in principle be distinguished from any other one, empirically, by means of measurements more or less like this one.[12]

And so it would seem that the right way to *find out* precisely where and precisely when collapses occur must be just to go out and *perform* these sorts of measurements.

The trouble is that, for a number of reasons, that turns out to be an extraordinarily difficult business to actually carry through.

The nicest one of those reasons has to do with the ways that measuring devices, in virtue of their macroscopicness, necessarily interact with their *environments*.

Let's see how that works.

Suppose that a black electron is on its way into a hardness measuring device and that the measuring device is in its ready state. And suppose that (as in figure 5.1) there happens to be a *molecule of air* sitting just to the right of the pointer, so that if the pointer were to swing to its "hard" position then the molecule would get pushed to the middle of the dial, and if the pointer were to swing

---

11. This can be confirmed by writing out the state in (5.2) in terms of eigenstates of color (for the electron) and of zip (for the pointer).

12. This is the sort of measurable property referred to in note 2 of Chapter 4.

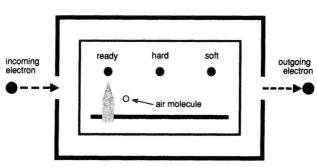

Figure 5.1

to its "soft" position then the molecule would get pushed to the right end of the dial.[13]

And now the electron passes through the hardness device, which swings the pointer, which pushes the air molecule (which behaves somewhat like a device for measuring the *position* of that pointer here);[14] and the reader (who is by now well versed in calculations about the behaviors of measuring devices) will have no trouble in confirming that the state of the electron and the hardness device and the molecule, when that's all done, *on my friend's theory*, is going to be

(5.4) $\quad 1/\sqrt{2}(|hard\rangle_e|\text{"hard"}\rangle_m|center\rangle_a + |soft\rangle_e|\text{"soft"}\rangle_m|right\rangle_a)$

where the vectors labeled with subscript a are state vectors of the air molecule.

That's how things stand at this point in the story, on my friend's theory, in the presence of an air molecule. Note that this state

13. Note that we are attributing both a fairly well-defined velocity (rest) and a fairly well-defined position (near the pointer) to this molecule *at the same time* here. The mass of an air molecule is large enough that doing so need not conflict with the uncertainty relations.

14. That is: the molecule is situated in such a way as to bring about a *correlation* between its *own* final position (once the swinging is over with) and the final position of the *pointer*. The molecule is situated in such a way as to make it possible to *infer* the final position of the pointer *from* the final position of the molecule.

(unlike the state in (5.2), which is how things stand at this same point in the story, on my friend's theory, in the *absence* of the air molecule) has *no* definite value of zip − color.[15]

And the way that things stand on my *own* theory, at this point in the story, in the presence of an air molecule, is

(5.5) $|hard\rangle_e|\text{"hard"}\rangle_m|center\rangle_a$ (with probability ½)

$|soft\rangle_e|\text{"soft"}\rangle_m|right\rangle_a$ (with probability ½)

Of course, neither of the states in (5.5) has any definite value of zip − color either. And so (5.4) and (5.5) (unlike (5.2) and (5.1)) cannot be distinguished from one another by means of measurements of zip − color.

And so, in the presence of an air molecule (unlike in the absence of one), my theory of the collapse and my friend's theory of the collapse cannot be distinguished from one another by means of measurements of zip − color.

What's going on here is that the state in (5.4) is nonseparable between the electron and the measuring instrument *and* the air molecule. Any measurement by means of which (5.4) can possibly be distinguished from (5.5)—any measurement, that is, by means of which my friend's theory can possibly be distinguished from my theory, in the presence of an air molecule (and of course such measurements will still, in principle, *exist*)—must necessarily be a measurement of an observable of the composite system consisting of all *three* of those objects. And the measurement of *that* observable (whatever, precisely, it turns out to be) will patently be a more difficult matter than the measurement of zip − color is.

15. The way to confirm that, of course, is to write (5.4) out in terms of eigenstates of zip − color.
What's going on here, by the way, can be looked at as another special case of the very general phenomenon described in note 4 of Chapter 4. The air molecule acts here as a device for measuring the position of the pointer; and zip, and zip − color, are both incompatible with that position, and so the dynamical equations of motion themselves will entail that interaction of the air molecule with the pointer will *disrupt* the values (that is: it will *randomize* the values) of zip and zip − color.

And this is also patently not the end of the story. There will be other air molecules around too, in general, and there will be tiny specks of dust, and imperceptible rays of light, and an unmanageable multitude of other sorts of microscopic systems as well, and all of them together (in virtue of their sensitivities to the position of the pointer; in virtue, that is, of the fact that each of them can act more or less as a *measuring instrument* for the position of that pointer) will *fantastically* increase the complexity of the observables which need to be measured (which is to say: it will fantastically increase the difficulty of measuring the observables which need to be measured) in order to distinguish between my theory and my friend's theory.

And of course the business of avoiding these sorts of complications by means of perfectly isolating the pointer from any interactions whatsoever with (say) molecules of air, and rays of light will be fantastically complicated too.

The upshot of all this (that is: what these arguments *establish*) is that different conjectures about precisely where and precisely when collapses occur are the sorts of conjectures which (for all practical purposes; or, rather, for all *presently* practical purposes) cannot be empirically distinguished from one another.[16] And so apparently the

---

16. What these arguments establish, by the way, has been rather widely misunderstood in the physical literature. The main confusion dates back (I think) to Daneri, Loinger, and Prosperi, 1962, and has since been carried on by Gottfried, 1966 (and there are related confusions in the work of Peres, 1980, and Gell-Mann and Hartle, 1990), and it goes like this: what these arguments establish is that different conjectures about *whether or not* collapses *ever* occur are the sorts of conjectures which (for all practical purposes; or, rather, for all presently practical purposes) cannot be empirically distinguished from one another.

Let's rehearse (again) why that can't possibly be true. The point is just that if the standard way of thinking about what it means to be in a superposition is the *right* way of thinking about what it means to be in a superposition (which is what all those misunderstanders always suppose), then (as we saw in Chapter 4) what proves that there *are* such things in the world as collapses is just the fact that *measurements have outcomes!* And the fact that measurements *do* have outcomes (as opposed to facts about precisely *when* they have outcomes) isn't the sort of fact that we learn by means of the difficult sorts of experiments we've been talking

best we can do at present is to try to think of precisely where and precisely when collapses might *possibly* occur (precisely where and precisely when they might occur, that is, without contradicting what we *do* know to be true, as of now, by experiment).

And it turns out to be hard (as we're about to see) to do even that.

Trying to Cook Up a Theory

Let's start over.

Let's see if we can think of precisely what it is that we *want* from a theory of the collapse.

Here's a first try at that:

i. We want it to guarantee that *measurements* (whatever, precisely, that term turns out to mean) *always have outcomes;* we want it to guarantee (that is) that there can never be any such thing in the world as a superposition of "measuring that *A* is true" and "measuring that *B* is true."

---

about *here.* The fact that measurements have outcomes is the kind of thing that we learn by means of direct introspection, by means of merely knowing that there are matters of fact about what our beliefs are!*

Of course (and this is going to be important), if it should turn out that the standard way of thinking about superpositions *isn't* the right way of thinking about them, then all bets are going to be off. But of that more later.

*It isn't easy to imagine precisely how all of those (extremely distinguished) misunderstanders can possibly have gotten themselves so confused. The argument we've been talking about, after all, is an argument about the probabilities of certain measurements having certain outcomes; any argument about the probabilities of certain measurements having certain outcomes will (needless to say) *critically depend* on the assumption that that there *are* (sooner or later) such things as the outcomes of those experiments; and so any argument about the probabilities of certain experiments having certain outcomes (supposing that the standard way of thinking about superpositions is the right way of thinking about them; which is, as I mentioned above, what all those guys *did* suppose) will critically depend on the assumption that there *are* (sooner or later) such things as *collapses;* and so (supposing what those guys supposed) any argument (like the one we've been talking about) about the probabilities of certain experiments having certain outcomes can't *possibly* be an argument to the effect that there might *not* (for all we know) be any such things as collapses; and yet that's *just* the sort of an argument that all those guys (go figure!) *took* it to be.

ii. We want it to preserve the familiar statistical connections between the outcomes of those measurements and the wave functions of the measured systems just before those measurements. That is: we want it to entail, or we want it at least to be consistent with, principle D of Chapter 2.

iii. We want it to be consistent with everything which is experimentally known to be true of the dynamics of physical systems. We want it, for example, to be consistent with the fact that isolated microscopic physical systems have never yet been observed *not* to behave in accordance with the linear dynamical equations of motion, the fact that such systems, in other words, have never yet been observed to undergo collapses.

The difficulty is in being absolutely explicit about (i).

Let's see if we can figure out how to do that. Let's try to be very practical. Let's focus somewhat more carefully on the physical mechanisms whereby the outcomes of measurements are ultimately recorded in measuring instruments.

What jumps out at you right away is that those recordings are typically stored in the *positions* of things: the positions of the tips of pointers, say, or the positions of drops of ink on scrolls of paper, or of bits of pencil lead in experimental notebooks.

Let's try to run with that. Perhaps it will suffice for what we want (which is to guarantee that measurements always have outcomes) to cook up a theory (if we can manage to) which somehow entails that every macroscopic object (that is, say, every object big enough to see, or every object even remotely big enough to see, or something like that) always (or maybe only almost always) has some definite particular position.

Let's see if we can make up a theory that does that.

Suppose we were to conjecture that every elementary particle in the world[17] occasionally (that is: once in some very great while), as

---

17. That is: the indivisible elementary pointlike constituents (electrons, say, and quarks, and so on) out of which every material object in the world is presumed to be made.

The idea of such elementary particles, by the way (that is: the idea of them which we shall want to make use of here), is an explicitly *nonrelativistic* idea; and so the

a matter of physical law, ceases (for an instant) to evolve in accordance with the dynamical equations of motion and undergoes a collapse which leaves it in an eigenstate of position. The times at which these collapses occur are stipulated to be absolutely random: there is merely a fixed, small, lawlike *probability,* per unit time, for each particle, that that particle, in that time interval, will undergo one of these collapses; and the point in space onto which the wave functions of these particles collapse (when they *do* collapse) will be determined probabilistically by the conventional quantum-mechanical probability formula (the one in principle D of Chapter 2); and all this is stipulated to hold for each particle *separately* (the times at which different particles undergo collapses, for example, will in general be unrelated to one another).

Let's spell this out somewhat more carefully. Consider a particle whose state at a certain instant is

$$(5.6) \qquad |A\rangle = a_1|x_1\rangle + a_2|x_2\rangle + a_3|x_3\rangle + \ldots$$

There is (according to this conjecture) a certain fixed, extremely small probability that this particle will undergo a collapse within the next, say, millisecond; and if it *does* happen to undergo one of those collapses, just at the moment when its state happens to be the one in (5.6), then the *probability* that that collapse will leave it in the state $|x_1\rangle$ will be $|a_1|^2$, and the probability that that collapse will leave it in the state $|x_2\rangle$ will be $|a_2|^2$, and so on.

Let's put it slightly differently. What happens in a state like (5.6), if a collapse happens to take place, is that one of the terms in (5.6) gets multiplied by a finite number, and all the rest get multiplied by 0; and then, until the next collapse takes place (which will likely be a very long time), the state of the particle evolves again in accordance with the equations of motion. The probability, if a collapse *does* occur, that the ith term in (5.6) is the one that gets multiplied by a finite number is $|a_i|^2$; and the *number* it gets multi-

theory of collapses which follows is going to be an explicitly nonrelativistic *theory.* The *relativization* of this theory is something we can begin to think about, if we're still in the mood, once the *non*relativistic theory is in place.

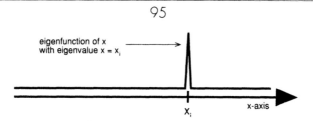

Figure 5.2

plied *by* (*if* it turns out to be the one that gets multiplied by a finite number) is $1/a_i$.

Let's put it one more way (this one will be particularly useful in what follows). What happens when a particle undergoes a collapse is that the *wave function* of the particle gets multiplied by an eigenfunction of the *position operator*, like the one depicted in figure 5.2; and the *probability* that the position *eigenvalue* of that position eigenfunction is x_i is stipulated to be equal to $|\langle x_i|w\rangle|^2$ (where $|w\rangle$ is the state of the particle at the moment just before the collapse occurs); and note that the outcome of this multiplication (that is: the product of these two wave functions), whatever $|w\rangle$ happens to be, is invariably *also* an eigenfunction of the position operator with eigenvalue x_i.[18]

Suppose that the probability per minute, for each particle, of undergoing a collapse is stipulated to be very, very, *very* small (one in trillions, say). Then the probability of our ever experimentally *observing* a collapse in an isolated microscopic system, a system (that is) which consists of small numbers of particles, will be very small too (even though such collapses will necessarily sometimes occur).

But (here's the punch line) consider what happens in *macroscopic* systems; consider what happens, for example, in *measuring instru-*

18. Note, by the way, that what I mean by *multiplying* two wave functions by one another, or taking the *product* of two wave functions, is something entirely different from taking the product of their two *state vectors*. The product of two state vectors is invariably a number. The product of two *wave functions*, on the other hand (as I'm using the term here), is defined to be that *wave function* whose value at each particular point in space is the product of the values of the two *multiplied* wave functions at that particular point in space.

ments. Think about the pointer in the hardness measuring device. It consists of trillions of particles (let's call them "1" and "2" and so on). The state in (5.2), for example, written out in terms of the states of those constituent particles, will look something like this:

$$(5.7) \qquad 1/\sqrt{2}(|x_1\rangle_1|x_1\rangle_2|x_1\rangle_3 \ldots)|\text{hard}\rangle_e + 1/\sqrt{2}(|x_2\rangle_1|x_2\rangle_2|x_2\rangle_3 \ldots)|\text{soft}\rangle_e$$

where x_1 is the position of the pointer when it's pointing to "hard" and x_2 is the position of the pointer when it's pointing to "soft."[19]

Suppose that one of the particles in the pointer, any one of them, when (5.7) obtains, were suddenly to undergo a collapse (and note that the probability of *that* occurring, within even a very small fraction of a second, will be very *high;* since the pointer consists of such a gigantic number of individual particles). Consider what that will do to the state in (5.7). Suppose that it happens to be the *i*th particle in the pointer that undergoes the collapse, and suppose (for example) that that particle happens to get collapsed onto the state $|x_1\rangle_i$. What that means is that the $|x_2\rangle_i$ vector in (5.7) gets multiplied by 0; and what *that* means (since $|x_2\rangle_i$ multiplies everything *else* in the second term in (5.7)) is that when that particle undergoes that collapse, if the collapse happens to be onto $|x_1\rangle_i$ (and the probability of that is of course precisely ½), then the entire second term in (5.7) (in (5.2), that is) will vanish, and the state will turn into the first of the states in (5.1).

And so the simple and beautiful conjecture which we are now entertaining (which is originally due to Ghirardi, Rimini, and Weber, 1986, and which was later put in a particularly nice form by Bell, 1987a) entails that collapses almost never happen to iso-

---

19. The fact that pointers on measuring devices typically consist of trillions of particles, by the way, is yet another of the reasons for the extraordinary difficulty of distinguishing, by means of experiments, between various different conjectures about precisely when collapses occur. The state in (5.7) (that is: the state in (5.2)) is non separable between the electron and *every one* of the constituents of the pointer; and so observables like zip – color, which distinguish between (5.2) and (5.1), must necessarily (even if the device and the electron were to be perfectly isolated from all of the rest of the world) be fantastically complicated ones; they must be observables which make reference to every individual one of all of those trillions of constituents.

lated microscopic systems. It *also* entails that states like the one in (5.2) will, almost certainly and almost *immediately*, collapse, with the standard quantum-mechanical probabilities, onto one or the other of the two states in (5.1). And all of that is of course precisely what we *want* from a theory of the collapse of the wave function; and all of it follows from a theory which can be formulated with perfect scientific explicitness, with no talk whatever (at a fundamental level) about "measurements" or "amplifications" or "recordings" or "observers" or "minds." And so now we really seem to be getting somewhere.

One technical difficulty needs to be attended to: The collapses described above leave the particles which undergo them in perfect eigenstates of the position operator, and of course that entails that the *momenta* and the *energies* of those particles (whatever their values may have been just *prior* to those collapses) will be completely uncertain just following those collapses, and *that* will give rise to a host of problems: The momenta which electrons in atoms might sometimes acquire in the course of such collapses, for example, would be enough to knock them right out of their orbits; and the energies which certain of the molecules of a gas might sometimes acquire in the course of such collapses would be enough to spontaneously *heat those gasses up*, and those sorts of things are experimentally known *not to occur.*[20]

The way to deal with that (and this is precisely what Ghirardi, Rimini, and Weber did) is to change the prescription slightly: Stipulate that when a particle undergoes a collapse, what its wave function gets multiplied by *isn't* an eigenstate of the position operator but is rather a *bell-shaped* function like the one in figure 5.3. Also stipulate that the *probability* of that bell curve's being centered at the point x_i (if such a collapse happens to occur) is proportional to $|\langle B_i | w \rangle|^2$ (where $|B_i\rangle$ is the bell-curve state centered at x_i and $|w\rangle$ is the state of the particle just *prior* to the collapse).

And note that in typical cases (when the wave function of the

20. That is: they're known not to occur as *often* as would be required, statistically, by this prescription.

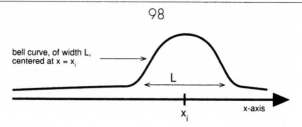

bell curve, of width L, centered at x = x$_i$

L

x$_i$

x-axis

Figure 5.3

particle just prior to the collapse is a good deal more spread out than the bell curve, and a good deal more smoothly varying than the bell curve) the multiplication of the initial wave function by a bell curve like the one in figure 5.3 will produce a *new* wave function which is *itself* much like the bell curve in figure 5.3 and which is centered at precisely the same point.

Let's see why that helps. Remember what it is that we want from collapses. What we want from them (what will suffice, at any rate) is to make the macroscopic world look as it looks to *us:* what we want from them is to insure that macroscopic objects invariably have determinate locations, or that they almost invariably have determinate locations, or at least that they almost invariably have *almost* determinate locations. And it turns out that this revised prescription can deliver that; it turns out that the bell curves can be made narrow enough so that whatever uncertainties there are in the positions of macroscopic things are almost invariably *microscopic* ones. And it turns out (and this is the punch line) that these curves can nonetheless be made *wide* enough (at the same time) so that the violations of the conservation of energy and of momentum which the multiplications by these curves will produce will be *too small to be observed.*[21]

But as a matter of fact this revised prescription gives rise to another sort of difficulty, a somewhat more subtle one.

21. Let's step back here a minute and look at precisely what sort of a theory we have.

This theory introduces two new fundamental constants of nature into our picture of the world: the probability, per unit time, per particle, of a collapse, and the

Consider again precisely what we want from a collapse; consider in precisely what *sense* we want it to guarantee that macroscopic objects almost always have "almost determinate locations." What we've found out is that the so-called widths of the bell curves of this prescription (that is: the length L in figure 5.3) can be made microscopic; but consider whether that's what *counts*, consider whether that will completely *suffice*.

The trouble is that the values of the bell-curve functions don't ever reach precisely 0, no matter how far away from their centers you get; and so the states that particles get left in, if they undergo collapses, on the revised prescription, are still (strictly speaking) *superpositions* of being *all over the place*; and so as a matter of fact, these collapses do *not* have the effect, on the standard way of thinking, of putting *anything* in an even approximately determinate position; and so this revised prescription (on the standard way of thinking) apparently *cannot* insure (for example) that experiments with measuring devices with pointers on them ever have outcomes after all.

The effect that these collapses *do* have, of course, is to put the state vectors of pointers *close* to *other* state vectors of those pointers in which those pointers *do* have (approximately) determinate positions. What needs to be made clear (if we want to stick with this theory) is how *that* does us any good. And it isn't easy to see (so far as I know) precisely how to do that.[22]

But let's suppose that there is a way to do that, and try to press on.

---

widths (the L's of figure 5.3) of the multiplying bell curves. The values of those constants, as well as the fact that such collapses occur at all, aren't things that Ghirardi, Rimini, and Weber ever make any attempt to explain: all of that is simply taken to be a part of the basic laws of nature.

This may strike some readers as unpleasantly ad hoc.

Those readers certainly have their nerve. What were they *expecting*, precisely? Let them go and reflect on what came *before*; let them go and reflect how much has been accomplished here.

22. Doing that (if there *is* a way of doing that) will apparently require some sort of a modification of the standard way of thinking.

Experiments with Television Screens

What we've been supposing so far is that every measuring instrument must necessarily include some sort of *pointer*, which indicates the outcome of the measurement. And we've been supposing that that pointer (if this instrument really deserves to be called a measuring instrument) must necessarily be a macroscopic physical object. And we've been supposing that that pointer must necessarily assume macroscopically different *spatial positions* in order to indicate different such outcomes. And it turns out that if all that is true, then the GRW theory (which is what we'll call it from now on) can do (i) and (ii) and probably (iii) as well.

The question, of course, is going to be whether all measuring instruments (or, rather, whether all reasonably *imaginable* measuring instruments) really do work like that.

Here's a standard sort of arrangement for measuring the hardness a particle: The particle gets fed into a hardness box like the one in figure 5.4, which separates hard particles from soft ones by means of magnetic fields. The incoming hard particles get directed (by the magnetic fields) toward point A on a TV screen, and the incoming soft particles get directed toward point B. The TV screen works like this: A particle striking the screen at (say) point B knocks

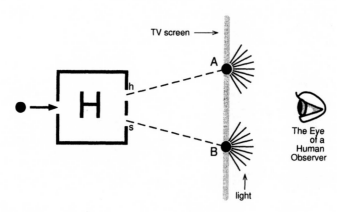

Figure 5.4

certain electrons in certain of the fluorescent atoms in the screen in the vicinity of point B into their so-called excited orbits (that is: it knocks those electrons into orbits around their atomic nuclei which are a good deal more *energetic* and a good deal more *unstable* than the orbits they're normally in); and soon thereafter those electrons spontaneously de-excite back to their original (stable, low-energy) orbits; and in the process of de-exciting (that is: in the process of losing their excess energy) they emit photons; and thus the vicinity of B becomes a luminous dot, which can be observed directly (that is: without any further artificial apparatus) by a human experimenter.

We want to inquire whether or not the GRW theory entails that a hardness measurement like that, on a particle which is initially (say) black, has an outcome. That will depend on whether or not there ever necessarily comes a time, in the course of measurement, when the position of a macroscopic object, or the positions of some gigantic collection of microscopic objects, is *correlated* with the hardness of that particle (let's call it P). With all that in mind, let's rehearse the stages of the measuring process (what follows here gets spelled out in a bit more detail in Albert and Vaidman, 1988) again.

First, the wave function of P gets magnetically separated (by the hardness box) into hard and soft components. No outcome of the hardness measurement (no collapse, that is) will be precipitated by that separation, since, as yet, nothing in the world save the position of P (nothing, that is, save a single, microscopic degree of freedom) is correlated to the hardness. Let's keep looking.

Next, P hits the screen, and at that stage the electrons in the fluorescent atoms get involved. Consider, however, whether those electrons get involved in such a way as to precipitate (via GRW) an outcome of the hardness measurement. Here's the crucial point: the GRW "collapses" are invariably collapses onto (nearly) eigenstates of position, but it's the *energies* of the fluorescent electrons, and *not* their positions, that get correlated, here, to the hardness of P! The GRW collapses aren't the right *sorts* of collapses to precipitate an outcome of the hardness measurement here.

Let's make this point somewhat more precise. If the initial state of P is $|black\rangle_P$, then, just after the impact of P on the TV screen, the (nonseparable) state of P and of the various fluorescent electrons in the vicinities of A and B will look (approximately; ideally) like this:

$$(5.8) \quad \tfrac{1}{\sqrt{2}} \, |hard, X = A\rangle_P |ex\rangle_{e1} \ldots |ex\rangle_{eN} |unex\rangle_{eN+1} \ldots |unex\rangle_{e2N}$$
$$+ \, \tfrac{1}{\sqrt{2}} \, |soft, X = B\rangle_P |unex\rangle_{e1} \ldots |unex\rangle_{eN} |ex\rangle_{eN+1}$$
$$\ldots |ex\rangle_{e2N}$$

where $e1 \ldots eN$ are fluorescent electrons in the vicinity of A, $eN + 1 \ldots e2N$ are fluorescent electrons in the vicinity of B, $|ex\rangle$ represents a state of being in an "excited" orbit, and $|unex\rangle$ represents a state of being in an "unexcited" orbit. Suppose, now, that a GRW "collapse" (that is, a multiplication of the wave function of one of the fluorescent electrons by a bell curve like the one depicted in figure 5.3) occurs. Consider whether this sort of a collapse will make one of the terms in (5.8) go away, allowing only the other to propagate. The problem, once again, is that these aren't the right sorts of collapses for that job, because $|ex\rangle$ can't be distinguished from $|unex\rangle$ in terms of the *position* of anything. Indeed, a GRW collapse will leave (5.8) almost entirely unchanged (except, perhaps, in the wave function of some single one of the many, many fluorescent electrons). And so no outcome of this measurement is going to emerge at this stage of things, either.[23]

We shall have to look still elsewhere. The next stage of the measuring process involves the decay of the excited electronic orbits and (in that process) the emission of photons. If the first term of (5.8) obtains, the photons would be emitted at A; if the second term obtains, the photons would be emitted at B. Those two states, then, *can* be distinguished, at least at the moment of emission, in terms of the *positions* of the *photons*. Now, so far, the GRW theory has been explicitly written down in a form which applies only to

23. I've left aside the whole question of the *probability* of a GRW collapse taking place at this stage of things (since, as we've just seen, a collapse like that won't do any good here anyway), but it ought to be noted in passing that that probability might well turn out to be extremely low. It's well known, after all, that the unaided human eye is capable of detecting very small numbers of photons; so perhaps only very small numbers of fluorescent electrons need, in principle, be involved here!

*non*relativistic quantum mechanics. The behaviors of *photons,* on the other hand (and particularly the *creation* of photons, by devices like TV screens), can be accounted for only in the context of a *relativistic quantum field theory.* It isn't completely clear as yet (for a number of rather technical reasons) how a GRW-type theory might treat *them.* If it turns out that photons can't experience GRW collapses, then of course no outcome of the hardness measurement can possibly emerge at this stage. But let's give the theory the benefit of the doubt: let's suppose that photons *can* experience GRW collapses. The problem at this stage of the measurement is going to be that that distinguishability is going to be extremely short-lived. It turns out that in almost no time, in far too little a time for a GRW collapse to be likely to occur (supposing that A and B are, say, a few centimeters apart, on a flat screen), the two photon wave functions described above will spread out (as shown in figure 5.5) so as to overlap almost entirely in position space, and the distinguishability in terms of positions will go away, and we shall be in just such a predicament as we found ourselves in at the previous stage of the measurement. No outcome, it seems, will emerge here, either.

But now we're running out of stages. The measurement (according to all the conventional wisdom about measurements) is already over! By now, after all, we have a recording; by now genuinely macroscopic changes (that is: changes which are thermodynamically irreversible, changes which are directly visible to the unaided human eye) have already taken place in the measuring apparatus.

And so it turns out that genuine recordings need not entail macroscopic changes in the position of anything; changes in (say) the energies of large numbers of atomic electrons (as in the above example) can be recordings too. It turns out (to put it slightly differently) that there can be genuinely macroscopic measuring instruments that (nonetheless) have absolutely no macroscopic *moving parts.*

That's what's been overlooked in the GRW proposal. What the GRW theory requires in order to produce an outcome isn't merely that the recording in the measuring apparatus be macroscopic (in any or all of the senses just described), but rather that the recording

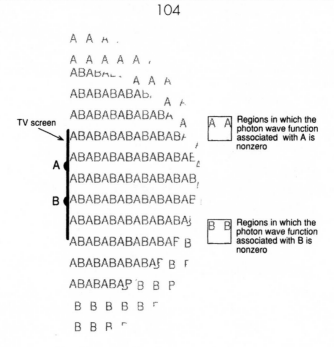

Figure 5.5

process involve macroscopic changes in the *position* of something. And the trouble is that no changes of that latter sort are involved in the kinds of measurements we've just been talking about.[24]

Inside the Observer's Head

Suppose, after all this, that we wanted to stick with the GRW theory anyway. What would that entail?

Well, we would have to deny that the measurement described above is over even once a recording exists. We would have to insist

24. Needless to say, this is not a point merely about measuring devices. The situation is as follows: if the GRW theory is the true physical theory of the world, then (astonishingly) things can in principle be cooked up in such a way as to guarantee that there literally fails to be any matter of fact about whether (say) Ralph Kramden's face or Ed Norton's face is the face that appears on some particular TV screen, at some particular moment.

(and certainly this is an ineluctable fact, when you come right down to it) that no measurement is absolutely over, no measurement absolutely requires an outcome, until there is a *sentient observer* who is actually *aware* of that outcome.

So, if we wanted to try to stick with this theory in spite of everything, the thing to do would be to insist that as a matter of fact we haven't run completely out of stages yet, and to go on looking, in those latter stages, for an outcome of this experiment (even though we've already looked right up to the retina of the observer and not found one); and of course the only place *left* to look at this point is going to be on the inside of the *nervous system* of the observer.

This is going to be an uncomfortable position to be in; what the possibility of entertaining some particular fundamental theory of the whole physical world is going to hinge on (if we go down this road) are the answers to certain detailed questions about the physiology of human beings; but let's try to press on and see how it might work. Here's one idea: Consider the two different physical states of the observer's retina (call them *RA* and *RB*) which arise in consequence of its being struck by light from one or the other of the two luminous spots on the fluorescent screen. Perhaps the retinal states *RA* and *RB* macroscopically differ from one another in (among other things) the *positions* of some gigantic collection of microscopic physical objects (the positions of some collection of ions, say). Whether or not that turns out to be so is of course a question for neurophysiology (and I presume that it's a question for the future of that subject), but the idea would be that if it *does* turn out like that, then the *observer's retina itself* can play the role of GRW's macroscopic pointer in this experiment: *it* (the retina) can bring about the collapse, *it* can suffice to finally precipitate an outcome.

Now of course it might turn out that *RA* and *RB* do *not* differ from one another in the positions of any sufficiently gigantic collection of physical objects. It might turn out, say, that the retina works more or less like a fluorescent screen and (consequently) that no outcome of this experiment can emerge at the stage of the retina, either. In that case we would presumably turn our attention next to the optic nerve, and *then*, finally (if things go badly with the optic nerve too), to the observer's brain itself.

If things were ever to get to that point, then the possibility of continuing to entertain the GRW theory would hinge directly on whether or not the brain state associated with seeing a luminous dot on the fluorescent screen at point A (call that *BA*) differs from the brain state associated with seeing a luminous dot on the fluorescent screen at point B (call that *BB*) in terms of the positions of any gigantic collection of physical components of the observer's visual cortex. If those two brain states *do* differ in that way (and that, once again, will be a question for neurophysiology), then the GRW theory *could* continue to be entertained.

But of course if it were to emerge, after all this, that *BA* and *BB* do not differ in precisely that way, if it were to emerge that they differ only in ways other than that, then the game would be over; then (that is) we would be absolutely out of stages.

Suppose that the human neurophysiology works out O.K.; suppose (that is) that it turns out that the human brain states *BA* and *BB* which I just described (the states which correspond to seeing a luminous dot at point A and seeing a luminous dot at point B) differ by, among other things, the *positions* of some gigantic number of ions.[25] How would things stand then?

Well, that would be a relief. That would mean that the GRW theory entails that at the point when the visual cortex of the experimenter gets into the game, then (and *only* then) an outcome of this experiment does finally emerge. Of course this outcome comes astonishingly late (later than we can possibly be comfortable with, later than anyone could ever have imagined). But suppose we're willing to swallow all that; suppose (that is) that we're willing to entertain the possibility that our intuitions can be radically false even about the absolutely familiar macroscopic external world. Then it would have to be admitted that (if you go strictly by the rule book) this outcome comes on time.

25. As a matter of fact, it has recently been argued (Aicardi, Borsellino, Ghirardi, and Grassi, 1991) that the brain states produced by different visual stimuli *are* likely to differ in that way. It isn't clear yet, however, what the story is with other sorts of sensory stimuli.

But what if at that point we were to begin to worry about the possibility of there being *other* sorts of sentient observers in the world (dolphins, say, or Martians or androids or whatever) for whom BA and BB *don't* differ in the appropriate way?

Or what if we were to begin to worry about the possibility of the development of (say) surgical techniques whereby ordinary human beings could be *transformed* into beings for whom BA and BB don't differ in the appropriate way?

Let's work through a science-fiction story (a story which, however, is everywhere scrupulously consistent with the hypothesis that the GRW theory is the true and complete theory of the world) about that last possibility.

First we'll need to set things up. The story involves a device (figure 5.6) for producing a correlation between the position of a certain microscopic particle P and the hardness of an electron; a sort of measuring device for hardness, in which the "pointer" is this particle called P. Here's how the device works: If P starts out in its middle position, and if a hard electron is fed through the device, then the hardness of that electron is unaffected by its passage through the device, and the device is unaffected too, except that P ends up, once the electron has passed all the way through, in its upper position; and if P starts in its middle position, and if a *soft* electron is fed through the device, then the hardness of that electron is unaffected by its passage through the device, and the device is unaffected too, except that P ends up in its *lower* position.

Suppose now that sometime in the distant future somebody named John undergoes a technically astonishing neurosurgical pro-

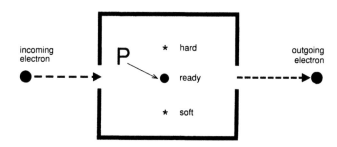

Figure 5.6

cedure which leaves him looking like he does in figure 5.7: John has a little door on either side of his head, and a device like the one I just described is now sitting in the middle of his brain, and (here comes the crucial point) the particular way in which that device is now hooked up to the rest of John's nervous system makes John behave as if his *occurrent beliefs* about the hardnesses of electrons which happen to pass through that device are determined *directly* by the position of *P*.

Here's what I mean. Suppose that John is presented with a hard electron and is requested to ascertain what the value of the hardness of that electron *is*. What he does is to take the electron into his head through his right door, and pass it through his surgically implanted device (with *P* initially in its middle position) and then expel it from his head through his left door. And when that's all done (when *P* is in its upper position but when, as yet, the value of

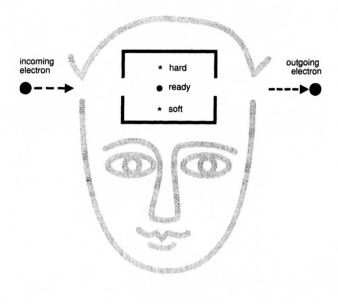

Figure 5.7

the hardness of the electron isn't recorded anywhere in John's brain *other* than in that position of *P*), John announces that he is, at present, consciously aware (as vividly and as completely as he is now aware of anything, or has ever been aware of anything in his life, he swears) of what the value of the hardness of the electron is, and that he would be delighted to tell *us* what that value is, if we would like to know.[26]

Let's think about what John says. There's going to be a temptation to suppose that John must somehow be mistaken, to say: "Look, how can it *be* that John is now consciously aware of what the value of the hardness of the electron is? Nothing in John's brain, other than the position of *P*, is *correlated* to the hardness of the electron now; and *P* isn't a natural part of John's brain *at all*; *P* is just a *surgical implant*, *P* is (after all) just a *particle!*" And imagine that John will say: "Look; whether or not *P* happens to be one of the components of my brain that I was born with is completely beside the point; I tell you now, from my own introspective experience, that if (as you tell me) nothing is now correlated to the hardness of that electron other than the position of *P*, then things must now be wired up in such a way that the position of *P* itself directly determines my present occurrent conscious beliefs about the value of the hardness of that electron!"

And John is now indeed in a position to announce, correctly, what the value of the hardness of that electron is, if he's asked to; and it's pretty clear that John can in principle be wired up so as to reproduce, in his present state, *any* of the behaviors of a genuine "knower" of that hardness *whatsoever;*[27] and John will claim (and who will we be to argue with him?) that, after all, no one is better

26. Of course, in the event that John actually *does* tell us that value, then (in the course of the physical process of telling us that) certain physical observables of John's brain other than the position of *P* would, no doubt, become correlated with the hardness of the measured electron. The point is that at *this* stage of things (when it's already the case that John claims to *know* the value of that hardness) they *aren't.*

27. And it's also pretty clear that John can in principle be wired up in such a way as to guarantee that (in his present state) he satisfies any *functional* criterion you can dream up for being a genuine "knower" of the hardness.

qualified than he himself to judge whether his own psychological experiences are really "genuine" ones or not!

Anyway, this much is certain: Either John is radically mistaken about what his own psychological experiences are, or else (if he isn't mistaken) nothing like a GRW theory can possibly insure that there are invariably determinate matters of fact about the psychological experiences of sentient observers at all. The point is that nothing macroscopic has happened in this story yet; and so if John now actually *has* an occurrent belief about the hardness of the electron, as he says he does, and as everything that can be empirically found out about him testifies he does, then John can indeed come to have such beliefs without anything macroscopic happening; and so (in the event that, say, the electron whose hardness John measures starts out in an eigenstate of *color*) nothing along the lines of a GRW theory is going to be able preclude the development of a *superposition* of states corresponding to *different* such beliefs.

And as a matter of fact, if John *can* come to have beliefs without anything macroscopic happening (which he says he can do, and which it seems he can do), then *no* theory of the collapse of the wave function *whatsoever* is going to be able to preclude the development of superpositions like that; because precluding superpositions like *that* will require that the collapse be inserted at a level (the level of *isolated microscopic systems*) at which we know, by experiment, that no collapses ever occur.[28]

Of course (on top of everything we've just been talking about) there hasn't ever been so much as a shred of what you might call *normal*

---

28. Perhaps this ought to be fleshed out a bit. Suppose that we were able to cook up a theory of the collapse which is patently as good as any theory of the collapse could imaginably be. Suppose, that is, that we were able to cook up a theory which is consistent with everything we know to be true of the behaviors of isolated microscopic systems and which entails (somehow) that superpositions of states which differ from one another in terms of anything macroscopic whatever (not just in terms of the *positions* of any gigantic number of particles) don't last long. And suppose that it were clear that this theory can indeed guarantee that an experiment carried out by an ordinary human observer invariably has an outcome.

The point is that a theory like that would nonetheless obviously fail to guarantee that for *John* in exactly the same way as the GRW theory does.

experimental evidence that the quantum state of any isolated physical system in the world ever fails to evolve in perfect accordance with the linear dynamical equations of motion.[29]

And so there seem to be a number of good reasons for looking for a different angle on this whole business.

29. That is: there hasn't been a shred of evidence that such failures ever take place, aside from the outcomes of certain *introspective* experiments that we carry out on *ourselves*.

The Dynamics by Itself

What It Feels Like to Be in a Superposition

The trouble with the quantum-mechanical equations of motion, according to Chapter 4 (and according to the conventional wisdom), runs as follows: The equations of motion (if they apply to everything) entail that in the event that somebody measures (say) the color of a hard electron, then the state of the measured electron (e) and the measuring device (m) and the human experimenter (h), when the experiment is over, will be:

(6.1) $1/\sqrt{2}$(|believes e black$\rangle_h$|"black"$\rangle_m$|black$\rangle_e$

              + |believes e white$\rangle_h$|"white"$\rangle_m$|white$\rangle_e$)

and of course, the state in (6.1) is (on the standard way of thinking) a state in which there is no matter of fact about what the color of the electron is, or about what the measuring device indicates its color to be, or even about what the experimenter takes its color to be; and the trouble with *that* (according to Chapter 4) is that we know, with certainty, by means of direct introspection, that there *is* a matter of fact about what we take the color of an electron like that to be, once we're all done measuring it; and so the state in (6.1) can't possibly be the way that experiments like that end up; and so the linear equations of motion must (in some instances, at least) be false.

But there's a small tradition of *resistance* to that conventional wisdom, which goes back to the late Hugh Everett III, who announced, in 1957, in a paper which is both extraordinarily suggestive and extraordinarily hard to understand, that he had discovered a way of coherently entertaining the possibility that the linear quantum-mechanical equations of motion are indeed (notwithstanding the argument rehearsed above) the true and complete equations of motion of the whole world. And that tradition merits some of our attention here.

What Everett announced (to put it a little more concretely) was that he had discovered some means of coherently entertaining the possibility that the states of things at the conclusions of color measurements of initially hard electrons really *are* (precisely as the linear equations of motion demand) superpositions like the one in (6.1); and the idea of what has become the canonical *interpretation* of Everett's paper (see, for example, DeWitt, 1970) is that the means of coherently entertaining that possibility that Everett must have had in mind (or perhaps the one which he *ought* to have had in mind) is to take the two components of a state like the one in (6.1) to represent (literally!) two physical *worlds*. The idea is that in the course of a measurement of (say) the color of a hard electron (the sort of measurement, that is, that leads to the state in (6.1)) the number of physical worlds there are literally increases from one to two, and that in each one of those worlds that color measurement actually has an outcome and the observer actually has a determinate belief about that outcome, and that those worlds are subsequently absolutely unaware of one another.

That's interesting. But there's a sense in which it can't be the whole story; there's a sense (that is) in which the above sort of talk, as it stands, isn't *well defined*. The trouble is that what worlds there are, at any particular instant, on this way of talking, will depend on *what separate terms there are in the universal state vector* at that instant; and what separate terms there are in that state vector at that particular instant will depend on what *basis* we choose to *write that vector down in;*[1] and of course there isn't anything in the

1. Here's what I mean: Suppose that a state like the one in (6.1) obtains; and consider (on this way of talking) what *worlds* that means that there presently are. What we're *tempted* to say (under these circumstances, on this way of talking) is that when (6.1) obtains there's one world in which the electron is black and there's

quantum-mechanical formalism itself which will pick any particular such basis out as the (somehow) *right* one to write things down in;[2] and so, if there's going to be any objective matter of fact about what worlds there are, at any given instant, on the many-worlds way of talking, then some new general principle is going to have to be *added* to the formalism which *does* pick out some particular basis as the right one to write things down in.

And of course the kind of principle we'll need is one that can guarantee that the worlds that come into being in the course of anything that can count as a measurement are all worlds in which there's a matter of fact about how that measurement *comes out.*

And so figuring out exactly what that principle ought to be is going to amount to figuring out exactly which *physical variables* there need to be matters of fact about, in order for there to be matters of fact about how measurements come out. And of course the business of figuring *that* out (as we discovered in the course of trying to cook up postulates of collapse) turns out to be very difficult.

And anyway, there's a more serious problem. There's a puzzle, when you talk like this, about what it could possibly mean to say (for example) that in the event that h carries out a color measure-

---

another world in which it's white. But note that the state in (6.1) could *also* have been written down like this:

$$\tfrac{1}{\sqrt{2}}(|Q+\rangle_{h\,+\,m}|\text{hard}\rangle_e + |Q-\rangle_{h\,+\,m}|\text{soft}\rangle_e)$$

where

$$|Q+\rangle_{h\,+\,m} = \tfrac{1}{\sqrt{2}}(|\text{believes } e \text{ black}\rangle_h|\text{``black''}\rangle_m + |\text{believes } e \text{ white}\rangle_h|\text{``white''}\rangle_m)$$

and

$$|Q-\rangle_{h\,+\,m} = \tfrac{1}{\sqrt{2}}(|\text{believes } e \text{ black}\rangle_h|\text{``black''}\rangle_m - |\text{believes } e \text{ white}\rangle_h|\text{``white''}\rangle_m)$$

and *that* makes things look (on this way of talking) as if there's one world in which the electron is *hard* and there's another in which it's *soft!*

2. On the standard quantum-mechanical formalism, after all, the choice of a set of basis vectors in which to write states down has no *physical* significance *whatsoever.*

ment of a hard electron, the "probability" that that measurement will come out white is ½. The trouble is that that sort of a measurement (on *this* way of talking) will *with certainty* give rise to *two worlds,* in one of which there's an *h* who sees that the outcome of the measurement is white, and in the other of which there's an *h* who sees that the outcome of the measurement is black; and there isn't going to be any matter of fact about which one of those two worlds is the *real* one, or about which one of those two *h*'s is the *original h.*

And there are myriad other difficulties with talking the many-worlds talk too (see, for example, Barrett, 1992), but I guess we need not rehearse any more of them here.

There are ways of making many-worlds talk sound less vulgar (which is to say: there are ways of making it sound less literal). But they don't get at what the real problems are.

Sometimes it gets proposed (for example) that there is exactly one physical world but that (when states like (6.1) obtain) there are two incompatible *stories* about that world, or maybe about how *h* sees that world, which are both somehow simultaneously *true.*[3]

It seems to me that that's really hard to understand. But one of the things that's obvious about it is that it runs into exactly the same sort of puzzle about what *probabilities* mean as the many-worlds talk does. Suppose, for example, that an observer named *h* carries out a measurement (just like the one we talked about above) of the color of a hard electron. Try to figure out what it might mean to say of an experiment like that (if you try to talk like this) that the probability that its outcome will be black is ½. The trouble is that this sort of talk is going to entail, with certainty, that there are *two* stories about what happens in an experiment like that; and there isn't going to be any matter of fact about *which one* of those

---

3. The most interesting attempt I know of at talking like that is Michael Lockwood's, in *Mind, Brain, and the Quantum* (Lockwood, 1989). Lockwood tries harder than anybody else does (with the possible exception of Lockwood's colleague David Deutsch, whose ideas show up at a number of crucial points in *Mind, Brain, and the Quantum*) to think about what it means (that is: to think about what it's *like*) for there to be more than one true story, when a state like (6.1) obtains, about what *h*'s experience is.

stories is the *true* one, and there isn't going to be any matter of fact about which one of those stories is the one that's about *the original h*.

I think it turns out to be a good deal more interesting to read Everett in a rather different way.

Suppose that there is only one world, and suppose that there is only one full story about that world that's true, and suppose that the linear quantum-mechanical equations of motion are the true and complete equations of motion of the world, and suppose that the standard way of thinking about what is means to be in a superposition is the right way of thinking about what it means to be in a superposition, and consider the question of what it would *feel* like to *be* in a state like the one in (6.1) (that is: the question of what it would feel like to be *the experimenter* in a state like the one in (6.1)).

That question wasn't confronted in Chapter 4. There didn't seem to be much of a point (back then) in confronting it. What seemed important was just that whatever it might feel like to be in a state like the one in (6.1), it certainly would *not* feel like what *we* feel like when we're all done measuring the color of a hard electron.[4]

But (since it turns out not to be easy to cook up a good-looking theory of the collapse, and since it turns out that no theory of the collapse whatsoever is going to be able to preclude the development of states like the one in (6.1) in people who undergo the kind of brain surgery described at the end of Chapter 5, and since there aren't any normal experimental reasons for believing that there *are* any such things as collapses) things are different now.

Here's a way to get started:

Suppose that the linear quantum-mechanical equations of motion were invariably true and (consequently) that observers like the one described above frequently did end up, at the conclusions of color measurements, in states like the one in (6.1).

Let's see if we can figure out what those equations would entail

4. That is: what seemed important was just that it could be established (by means of the argument on page 112) that as a matter of fact human experimenters don't *end up* in states like that, at the conclusions of those sorts of measurements.

about how an observer like that, in a state like the one in (6.1), would *respond* to *questions* about how she feels (that is: about what her mental state is). Maybe that will tell us something.

The most obvious question to ask is: "What is your present belief about the color of the electron?" But that question turns out not to be of much use here. Here's why: Suppose that the observer in question (the one that's now in the state in (6.1)) gives honest responses to such questions; suppose, that is, that when her brain state is |believes *e* black⟩ she invariably responds to such a question by saying the word "black," and when her brain state is |believes *e* white⟩ she invariably responds to such a question by saying the word "white." The problem is that precisely the same linearity of the equations of motion which brought about the superposition of different brain states in the state in (6.1) in the first place will now entail that if we were to address this sort of a question to this sort of an observer, when (6.1) obtains, then the state of the world after she *responds* to the question will be a superposition of one in which she says "black" and another in which she says "white"; and of course it won't be any easier to interpret a "response" like that than it was to interpret the superposition of brain states in (6.1) that that response was intended to be a *description* of!

But there are other sorts of questions that turn out to be more informative.

Note, to begin with, that it follows from the linearity of the operators that represent observables of quantum-mechanical systems (the sort of linearity that was defined in equation (2.9)) that if any observable O of any quantum-mechanical system S has some particular determinate value in the state $|A\rangle_S$, and if O also has that same determinate value in some other state $|B\rangle_S$, then O will necessarily *also* have precisely that same determinate value in any linear *superposition* of those two states.[5]

---

5. That's an entirely commonsensical way for observables to behave, if you think it through. Suppose, for example, that there's a particle which is in a superposition of being located in the right half and in the left half of a certain box. What the linearity of the observables of a particle like that is going to entail (or rather, *one* of the things that it's going to entail) is that that particle is in an eigenstate of the observable "is the particle anywhere in the box at all?" with eigenvalue "yes."

Let's apply that to the superposition of states in (6.1).

Suppose that we were to say this to h: "Don't tell me whether you believe the electron to be black or you believe it to be white, but tell me merely whether or not *one* of those two is the case; tell me (in other words) merely whether or not you now *have* any particular definite belief (not uncertain and not confused and not vague and not superposed) about the value of the color of this electron."

Now, if (when we ask h that) the state $|$believes e black$\rangle_h \times |$"black"$\rangle_m|$black$\rangle_e$ obtains, and if h is indeed an honest and competent reporter of her mental states, then she will presumably answer, "Yes, I *have* some definite belief at present, one of those two *is* the case"; and of course she will answer in precisely the same way in the event that $|$believes e white$\rangle_h|$"white"$\rangle_m|$white$\rangle_e$ obtains.

And so responding to this particular question in this particular way (by saying "yes") is an observable property of h in both of those states, and consequently (and this is the punch line) it will also be an observable property of her in any superposition of those two brain states, and consequently (in particular) it will be an observable property of her in (6.1).

That's odd. Look what we've found out: On the one hand, the dynamical equations of motion predict that h is going to end up, at the conclusion of a measurement like the one we've been talking about, in the state in (6.1), and not in either one of the brain states associated with any definite particular belief about the color of the electron; on the other hand, we have just now discovered that those same equations also predict that when a state like (6.1) obtains, h is necessarily going to be convinced (or at any rate she is necessarily going to *report*) that she *does* have a definite particular belief about the color of the electron. And so when a state like (6.1) obtains, h is apparently going to be radically deceived even about what *her own occurrent mental state* is.

And so it turns out that there was a hell of a lot too much being taken for granted when we got convinced (back in Chapter 4) that there is some particular point in the course of the sort of measurement we've been talking about by which a collapse of the wave function must necessarily already have taken place, some particular point (that is) at which the dynamical equations of motion together with the standard way of thinking about what it means to be in a superposition somehow flatly contradicts what we unmistakably know to be true of our own mental lives.

Let's go on. Suppose that h carries out a measurement of the color of a hard of electron with a color measuring device called $m1$, and suppose that when that's done (that is: when a state like (6.1) obtains) h carries out a *second* measurement of the color of that electron, with a *second* color measuring device called $m2$. When *that's* all done, the state of h and of the two measuring devices and the electron (if the measuring devices are good, and if h is competent, and if everything evolves in accordance with the linear dynamical equations of motion) is going to look like this:

(6.2) $1/\sqrt{2}($ |believes outcome of first measurement is "black"
                    and believes outcome of second measurement is
                    "black"$\rangle_h$|"black"$\rangle_{m1}$|"black"$\rangle_{m2}$|black$\rangle_e$

        + |believes outcome of first measurement is "white"
              and believes outcome of second measurement is
              "white"$\rangle_h$|"white"$\rangle_{m1}$|"white"$\rangle_{m2}$|white$\rangle_e)$

And suppose that at that point (when (6.2) obtains) we were to say to h: "Don't tell me what the outcomes of either of those two color measurements were; just tell me whether or not you now believe that those two measurements both had definite outcomes, and whether or not those two outcomes were *the same*."

It will follow from the same sorts of arguments as we gave above that h's response to a question like that (even though, as a matter of fact, on the standard way of thinking, *neither* of those experiments had any definite outcome) will necessarily be: "Yes, they both had definite outcomes, and both of those outcomes were the same."

And it will also follow from the same sorts of arguments that if two observers were both to carry out measurements of the color of some particular initially hard electron, and if they were subsequently to talk to one another about the outcomes of their respective experiments (if they were both, that is, to *check up* on one another), then both of those observers will report, falsely, that the *other* observer has reported some definite particular outcome of *her* measurement, and both of them will report that that reported outcome is completely in agreement with her own.

Let's make up a name for all that. Let's say that when a state like (6.1) obtains, then (even though there isn't any matter of fact about what the color of the electron is, and even though there isn't even any matter of fact about what h's *belief* about the color of the electron is) what the dynamics entails is that h "effectively knows" what the color of the electron is.

Let's go on some more. Suppose that h is confronted with an infinite collection of electrons, all of which are initially hard, and that h undertakes to measure the color of *each one* of those electrons.

Before those measurements start, the state of h and of those electrons (whose names are 1, 2, . . .) and of h's color measuring devices (whose names are, respectively, $m1$, $m2$, . . .) is:

$$(6.3) \qquad |ready\rangle_h |ready\rangle_{m1} |hard\rangle_1 |ready\rangle_{m2} |hard\rangle_2 |ready\rangle_{m3} |hard\rangle_3 \ldots$$

Once the measurement of the color of electron 1 is done, the state is:

$$(6.4) \qquad 1/\sqrt{2}(|believes\ 1\ black\rangle_h |"black"\rangle_{m1} |black\rangle_1$$
$$+ |believes\ 1\ white\rangle_h |"white"\rangle_{m1} |white\rangle_1)$$
$$\times |ready\rangle_{m2} |hard\rangle_2 |ready\rangle_{m3} |hard\rangle_3 \ldots$$

And once the measurement of the color of electron 2 is done, the state is:

$$(6.5) \qquad 1/\sqrt{4}\{(|believes\ 1\ black\ and\ 2\ black\rangle_h |"black"\rangle_{m1} \times$$
$$|"black"\rangle_{m2} |black\rangle_1 |black\rangle_2)$$
$$+ (|believes\ 1\ black\ and\ 2\ white\rangle_h |"black"\rangle_{m1} \times$$
$$|"white"\rangle_{m2} |black\rangle_1 |white\rangle_2)$$
$$+ (|believes\ 1\ white\ and\ 2\ black\rangle_h |"white"\rangle_{m1} \times$$
$$|"black"\rangle_{m2} |white\rangle_1 |black\rangle_2)$$

+ (|believes 1 white and 2 white⟩$_h$|"white"⟩$_{m1}$ ×
|"white"⟩$_{m2}$|white⟩$_1$|white⟩$_2$)}

× |ready⟩$_{m3}$|hard⟩$_3$ . . .

And so on. The number of separate mathematical terms in the state vector of the world (if you write it out in the sort of basis that's used here) will increase geometrically (like the numbers of the branches in the diagram in figure 6.1, as you work your way up) as the number of color measurements increases.

Now, suppose that once the first N of those measurements are complete we say this to h: "Don't tell me what the color of electron 1 or electron 2 or any particular one of the first N electrons turned out to be; tell me merely whether or not you believe that each one of those electrons now has a definite color, and tell me also (if the answer to that first question is yes) what *fraction* of those first N electrons turned out to be *black*."

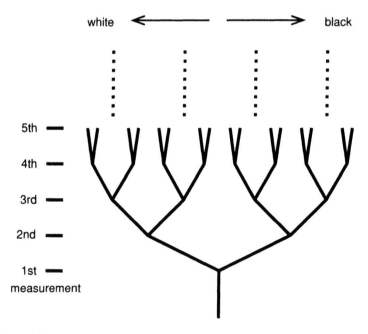

Figure 6.1

That won't tell us much, as it stands. The answer to the first question (as we've already seen) is going to be "yes" (and moreover, at that point, it's going to be a physical fact about the world that h *effectively knows* the color of each of those first N electrons). But of course h *isn't* going to produce any coherent answer to the *second* question; once h has responded to *that* question (if h is a competent converser on these matters), the state of the world is going to be a superposition of states (like the superposition that arises in the event that we ask h what the color of the electron is when (6.1) obtains) in which h answers that question in various different ways.

But here's something curious: It happens that in the limit as N goes to infinity (that is: in the limit as the number of color measurements which h has so far performed goes to infinity), the state of the world will, with certainty, *approach* a state in which h *will* answer that question in a perfectly *determinate* way, and in which the answer h gives will with certainty be "$1/2$" (which is, of course, precisely what *ordinary* quantum-mechanics will predict, with certainty, about that response, in that limit).[6]

And that turns out to be an instance of something a good deal

6. It isn't hard to see why that sort of thing ought to be true. Here's how to start out: Consider (for example) an infinite collection of electrons (let's call them 1, 2, 3, . . .), all of which are in the |hard⟩ state; and consider the following *observable* of that collection: $O_N = (1/N) \times$ (the number of *black* electrons among the first N electrons).

Now, *ordinary* quantum mechanics (that is: quantum mechanics with a *collapse*) entails that if the color of each one of those electrons were to be measured, then (since there are infinitely many of those electrons in this collection) precisely half of those measurements would with certainty come out "black," and precisely half of them would with certainty come out "white." Moreover, O_N (for any value of N) is *compatible* with the colors of every one of those electrons. And so it follows that ordinary quantum mechanics entails that as N approaches infinity, the probability that a measurement of the value of O_N on the collection of electrons described above will find the value $1/2$ will approach 1. And so it follows that that collection of electrons (*prior* to any measurement) must be in an *eigenstate* of whatever operator it is that O_N approaches as N approaches infinity, with eigenvalue $1/2$.

And it will follow from all *that* that *whether or not* there are ever any such things as collapses, the state of a composite system consisting of that collection of initially

more general, which runs as follows: Suppose that an observer h is confronted with an infinite ensemble of identical systems in identical states and that she carries out a certain identical measurement on each one of them. Then, even though there will actually be no matter of fact about what h takes the outcomes of *any* of those measurements to be, nonetheless as the number of those measurements which have already been carried out goes to infinity, the state of the world will approach (as a well-defined mathematical limit) a state in which the reports of h about the *statistical frequency* of any particular outcome of those measurements will be perfectly definite, and also perfectly in accord with the standard *quantum-mechanical* predictions about what that frequency ought to be.

So it turns out not to be altogether impossible (even if the standard way of thinking about what it means to be in a superposition is the right way of thinking about it) that the state we end up in at the conclusion of a measurement of the color of a hard electron is the one in (6.1). And so everything we've been thinking about the measurement problem up till now isn't right.

And what all this obviously suggests is that maybe there just isn't any such thing as a measurement problem.

That is: maybe (even if the standard way of thinking about what it means to be in a superposition is the right way of thinking about what it means to be in a superposition) the linear dynamical laws are nonetheless the complete laws of the evolution of the *entire world,* and maybe all of the appearances to the contrary (like the appearance that experiments have outcomes, and the appearance that the world doesn't evolve deterministically) turn out to be just the sorts of *delusions* which *those laws themselves* can be shown to *bring on!*

---

hard electrons and of a competent *observer* who has just carried out measurements of the *colors* of all those electrons will, with certainty, be an eigenstate of that observer's *reporting* (if she's asked) that the value of whatever operator it is that O_N approaches as N approaches infinity is $\frac{1}{2}$.

A detailed mathematical discussion of all this stuff (with much nicer proofs than the one above) can be found in the doctoral dissertation of Jeff Barrett (1991).

This is an amazingly cool idea (let's call it "the bare theory"), and *this* is the idea that it strikes me as interesting to read into Everett's paper.[7]

Notwithstanding all the stuff we've just learned, however, it seems to me that the bare theory can't be quite right either.

Note, for example, that if the bare theory is true, then there will be matters of fact about what we think about (say) the frequencies of "black" outcomes of measurements of the color of hard electrons *only* (if at all) in the limit as the number of those measurements goes to infinity. And so, if the bare theory is true (and since only a finite number of such measurements has ever actually been carried out by any one of us, or even in the entire history of the world), then there can't now be any matter of fact (notwithstanding our delusion that there is one) about what we take those frequencies to *be*. And so, if the bare theory is true, then there can't be any matter of fact (notwithstanding our delusion that there is one) about whether or not we take those frequencies to be in accordance with the standard quantum-mechanical *predictions* about them. And so, if the bare theory is true, it isn't clear what sorts of reasons we can possibly have for *believing* in anything like quantum mechanics (which is what the bare theory is supposed to be a way of making *sense* of) in the *first* place.[8]

And as a matter of fact, if the bare theory is true, then it seems extraordinarily unlikely that the present quantum state of the world can possibly be one of those in which there's even a matter of fact about whether or not any sentient experimenters exist at all. And of course in the event that there *isn't* any matter of fact about

---

7. Of course, the hypothesis that the equations of motion are always exactly right is *also* the sort of thing that Daneri, Loinger, and Prosperi and all those other guys in note 16 of Chapter 5 took themselves to be adherents of.

The trouble is that (astonishingly) it never seems to have *occurred* to *those* guys that it follows from that hypothesis that experiments almost never have outcomes; and so none of them ever worried about how to come to terms with that; and so none of them ever entered into the sorts of considerations that we're in the midst of here.

8. This very nice way of putting the problem is due to Joshua Newman.

whether or not any sentient experimenters exist, then it becomes unintelligible even to inquire (as we've been doing here) about what sorts of things such experimenters will *report.*

And then (as far as I can tell) all bets are off.

And so it seems to me *not* to be entertainable, in any of the ways we've talked about so far, or in any *other* way I know of, that the linear quantum-mechanical equations of motion are the true and complete equations of motion of the whole world. And that's that.

But there are nonetheless interesting things to be learned about the measurement problem (things that it will be well to bear in mind in connection with the problems we ran into at the end of the last chapter, and in connection with problems we *will* run into at the end of the *next* chapter) in this stuff about what superpositions feel like.

What that stuff shows, I think, is that precisely that feature of those equations which makes it *clear* that they cannot possibly be the true and complete equations of motion of the whole world (that is: their *linearity*) *also* makes it radically *un*clear *how much* of the world and *which parts* of the world those equations possibly *can* be the true and complete equations of motion of.

What I think it shows (to put it another way) is that there can be *no such thing* as a definitive list of what there have absolutely got to be matters of fact about which is scientifically fit to serve as an "observational basis" from which all attempts at fixing quantum mechanics up must *start out.*

What I think it shows is that what there are and what there aren't determinate matters of fact about, even in connection with the most mundane and everyday macroscopic features of the external physical world, and even in connection with the most mundane and everyday features of *our own mental lives,* is something which we shall ultimately have to *learn* (in some part) from whatever turns out to be the *best* way of fixing quantum mechanics up.[9]

---

9. But note that that learning will be no straightforward matter, since one of the things that all this raises difficult questions about is the very business of *seeking out* the best way of fixing quantum mechanics up!

The Dynamics Almost by Itself

Let's start again.

Suppose that there's just one world. And suppose that there's just one complete story of the world that's true.

And suppose that quantum-mechanical state vectors are complete descriptions of physical systems. And suppose that the dynamical equations of motion are always exactly right.

And suppose that we should like to insist, as a matter of principle, that healthy people can correctly report whether or not they themselves have any determinate belief about (say) the position of some particular pointer.

Then (since the dynamical equations of motion entail that healthy people in superpositions of brain states corresponding to different beliefs about the position of some particular pointer will with certainty report that *they have some determinate belief* about the position of that pointer) there's going to have to be something funny about how mental states supervene on brain states.[10]

Let's see if we can cook up something funny like that.

Think of *h* when she's about to measure the color of the hard electron, when she's in her "ready" state. When the measurement is over, the physical state of *h* and her measuring device and the electron is going to be the one in (6.1). That's what's dictated, with certainty, by the deterministic equations of motion.

Suppose, however, that all that's true, but that the evolution of *h*'s *mental* state in the course of a measurement like this one is explicitly *probablistic*. Here's how things would go in this particular case: *h* starts out (with certainty) in the mental state associated with |ready⟩$_h$, and she ends up (with equal probabilities) either in the mental state associated with |believes *e* black⟩$_h$ or in the mental state associated with |believes *e* white⟩$_h$.[11] What's *certain* about how she ends up, though, is that she ends up (just as she testifies she

---

10. That is, it's going to have to be the case that somebody's believing that such-and-such is *not* identical with some particular state of that person's *brain* (the state we've been calling |believes such-and-such⟩) obtaining.

11. It's obvious how this ought to be generalized: In the event that the initial state of the electron is *a*|white⟩ + *b*|black⟩, and in the event that *h* measures that electron's color, then (as above) *h* will start out, with certainty, in the mental state

does) with some perfectly determinate belief about what the color of the electron is.[12]

So far so good. Let's try to take it a little further.

Whatever belief h *does* end up with, when (6.1) obtains, is necessarily going to be a false belief. But there are very natural ways of cooking things up so as to guarantee that that belief will nonetheless have an important kind of *effective validity*, at least in so far as h is concerned;[13] there are ways of cooking things up (that is) so as to guarantee that the future evolution of h's mental state will proceed, in general, exactly *as if h's beliefs were* true.

Here's what I mean.

Suppose that the mental state that h ends up in when (6.1) obtains (call that time t) happens to be the one associated with |believes e black$\rangle_h$, and suppose that she subsequently repeats that color measurement (with another color measuring device) on that same electron. When that's done, the physical state of things is going to be the one in (6.2), and h will with certainty (on this proposal) end up in the mental state associated with |believes outcome of first measurement is "black" and believes outcome of second measurement is "black"$\rangle_h$.[14] And that's precisely how h's mental state *would* have ended up, with certainty (on this proposal), in the event that her belief that the electron was *black* at time t (which was false) had been *true*.

And suppose (just as above) that a state like the one in (6.1)

---

associated with |ready$\rangle_h$, and she'll end up in the mental state associated with |believes e white$\rangle_h$ with probability $|a|^2$, and she'll end up in the mental state associated with |believes e black$\rangle_h$ with probability $|b|^2$.

12. This sort of thing was first suggested quite a long time ago, but for somewhat different reasons (for reasons which had nothing to do with what the equations of motion dictate about what it feels like to be in a superposition) by Bernard d'Espagnat (1971).

13. Of course, there was a sense in which it seemed right to say that h *effectively knows* what the color of the electron is, when (6.1) obtains, on the *bare* theory too. But what we're talking about now will amount to something a good deal stronger than that.

14. And of course in the event that h's mental state at t happens to be the one associated with |believes e white$\rangle_h$, then h will with certainty end up in the mental state associated with |believes outcome of first measurement is "white" and believes outcome of second measurement is "white"$\rangle_h$.

obtains, and suppose that h's mental state happens to be the one associated with |believes e black$\rangle_h$, and suppose that she subsequently carries out a measurement of the *hardness* of that same electron; then, when *that's* done, the physical state of things is going to be

(6.6) $1/\sqrt{4}$ {((|believes outcome of first measurement is "black" and believes outcome of second measurement is "hard"$\rangle_h$|"black"$\rangle_{m1}$|"hard"$\rangle_{m2}$|hard$\rangle_e$)

+ (|believes outcome of first measurement is "black" and believes outcome of second measurement is "soft"$\rangle_h$|"black"$\rangle_{m1}$|"soft"$\rangle_{m2}$|soft$\rangle_e$)

+ (|believes outcome of first measurement is "white" and believes outcome of second measurement is "hard"$\rangle_h$|"white"$\rangle_{m1}$|"hard"$\rangle_{m2}$|hard$\rangle_e$)

− (|believes outcome of first measurement is "white" and believes outcome of second measurement is "soft"$\rangle_h$|"white"$\rangle_{m1}$|"soft"$\rangle_{m2}$|soft$\rangle_e$)}

(where $m1$ is the color measuring device and $m2$ is the hardness measuring device), and the probability that h will end up in the mental state associated with |believes outcome of first measurement is "black" and believes outcome of second one is "hard"$\rangle_h$ will be $1/2$, and the probability that she will end up in the mental state associated with |believes outcome of first measurement is "black" and believes outcome of second measurement is "soft"$\rangle_h$ will be $1/2$, and the probability of her ending up in the mental states associated with either |believes outcome of first measurement is "white" and believes outcome of second measurement is "hard"$\rangle_h$ or |believes outcome of first measurement is "white" and believes outcome of second measurement is "soft"$\rangle_h$ will be 0.

And suppose that a state like the one in (6.4) obtains and that h's mental state happens to be the one associated with |believes 1 black$\rangle_h$, and suppose that h now carries out a measurement of the color of electron 2 (in which case the physical state of the world will become the one in (6.5)). Then (on this proposal) the probability of h's ending up in the mental state associated with |believes 1 black and 2 black⟩$_h$ will be ½, the probability of h's ending up in the mental state associated with |believes 1 black and 2 white⟩$_h$ will be ½, and the probabilities of h's ending up in the mental states associated with either |believes 1 white and 2 black⟩$_h$ or |believes 1 white and 2 white⟩$_h$ will both be 0.

And so on.[15]

That (technically) will do the trick. On this proposal, quantum-mechanical wave functions are complete descriptions of the physical states of things, and those wave functions invariably evolve in perfect accordance with the dynamical equations of motion, and it makes no physical difference at all what *basis* we choose to write those wave functions *down* in,[16] and measurements carried out by sentient observers (that is: by observers with *minds*) invariably have determinate *outcomes* in the minds of those observers, and the statistical *distributions* of those outcomes will be the usual quantum-mechanical ones, and there isn't anything mysterious about how *probabilities* come up in this theory,[17] and the *reports* of sentient observers about their own mental states will invariably, on this proposal, be *correct*.

15. What's been said so far (a slightly more detailed account of which, by the way, can be found in Albert and Loewer, 1988) doesn't amount to a completely general set of laws of the evolution of mental states; but laws like that can be cooked up, and they can be cooked up in such a way as to guarantee that everything I've said about them so far will be true.

16. Of course, there will (on this picture, and on every way of attempting to make sense of quantum mechanics) be some particular basis of brain states which correspond to (as it were) "eigenstates of mentality"; but what basis that *is* will by no means be a matter of conventional choice; what basis that is will entirely depend (rather) on the *physical structure* of the *brains* in question. The brain state that corresponds to believing that a certain electron is black, for example, will presumably be the one which (purely in virtue of the dynamical equations of motion) disposes its owner to respond to an utterance like "What color do you believe the electron to be?" with an utterance like "I believe the electron to be black." And of course what brain state *that* will be will be a completely basis-independent, straightforwardly physical question!

17. The way that probabilities come up in *this* theory, after all, is that they get *put* into it by *fiat;* and that fiat stipulates that those probabilities are to be understood in precisely the conventional way.

And of course this view of the world is a thoroughly realist one (that is: this view entails that there is invariably a single correct objective description of the entire physical and mental universe, even if nobody happens to know what that description *is*); and this view (even though it's an explicitly dualist view) entails that the mental parts of the world have no effects whatever on its physical parts (that is: this view isn't at all like any of the dualist theories of *collapse, this* view entails that the physical world is *causally closed*).

But the dualism of this sort of a picture is nonetheless pretty bad. On this proposal (for example) all but one of the terms in a superposition like the one in (6.1) represent (as it were) *mindless hulks;* and *which one* of those terms is *not* a mindless hulk can't be deduced from the physical state of the world, or from the outcome of any sort of an experiment; and it will follow from this proposal that most of the people we take ourselves to have met in our lives have as a matter of fact *been* such hulks, and not really people (not really animate, that is) at all!

Here's a way to partly fix *that* up:

Suppose that every sentient physical system there is is associated not with a single mind but rather with a *continuous infinity* of minds; and suppose (this is part of the proposal too) that the measure of the infinite subset of those minds which happen to be in some particular mental state at any particular time is equal to the square of the absolute value of the coefficient of the brain state associated with that mental state, in the wave function of the world, at that particular time (so that, for example, when states like (6.1) obtain, half of h's continuous infinity of minds will believe that the electron is black, and half of them will believe that the electron is white).

The time evolution of each individual mind, on this proposal, is precisely the probabilistic one described above (the one that we cooked up for the single-mind proposal), but since (on *this* proposal) there are always a continuous infinity of minds (or else no minds at all) in any particular mental state, the evolution of the minds of any particular sentient observer *as a set* is invariably (that is: with probability 1) going to be *deterministic*. Moreover, at any particular instant, the mental states of the minds of any particular observer will necessarily be distributed in accordance with the prescription of the last paragraph. So this proposal is going to entail that what you might call the "global" mental state of every sentient being *is* uniquely fixed by the physical state of the world.[18]

And there's something else about this kind of a picture that's nice: this kind of a picture is *local*. That's surprising. That's precisely the sort of thing that Bell's theorem was thought to have ruled out. Let's see how it works.

Consider an EPR-type state:

(6.7) $|black\rangle_1|white\rangle_2 - |white\rangle_1|black\rangle_2$

and suppose that electron 1 is located at point 1 and that electron 2 is located at point 2 and that an observer named $h1$ (located at point 1) measures some spin observable of electron 1 and that an observer named $h2$ (located at point 2) measures some spin observable (not necessarily the same one) of electron 2.

What Bell proved is that there can't be any local way of accounting for the observed correlations between the outcomes of measurements like that; but of course (and this is the crux of the whole business) the idea that there ever *are* matters of fact about the "outcomes" of a pair of measurements like that is just what *this* sort of a picture *denies!*

Let's go through it carefully.

At the conclusion of a pair of measurements like the one just described, on *this* picture, the state of the world is going to be a superposition of states, in each of which each of those two measurements have one or the other of their two different possible outcomes. And at that point, on this picture, no matter what spin observable of electron 1 gets measured by $h1$ and no matter what spin observable of electron 2 gets measured by $h2$, half of $h1$'s minds are going to believe that the outcome of whatever measurement *she* did was $+1$, and the other half of her minds are going to

18. This is the so-called many-minds interpretation of quantum mechanics, which was first proposed by Barry Loewer and myself (Albert and Loewer, 1988).

believe that the outcome of whatever measurement she did was -1, and half of $h2$'s minds are going to believe that the outcome of whatever measurement *she* did was $+1$, and the other half of her minds are going to believe that the outcome of whatever measurement she did was -1. (The reader will have no trouble in explicitly confirming all this, and in confirming that none of this depends on the *time order* in which the two measurements get carried out.)

And this is where things get a little more subtle.

What it's hard not to do, at first, at this point in the story, is to imagine that there are matters of fact (when this sort of a superposition obtains) about the degree to which the states of the minds of $h1$ and the states of the minds of $h2$ are *correlated* with one another.

But the thing is that there *aren't* any matters of fact about anything like that.

All that ever actually *happens* (insofar as any question of *correlations* is concerned) is that at the point (later on) when $h1$ actually *communicates* with $h2$ (and that communication is of course going to be mediated by some *local* interaction and governed by the *local* equations of motion), then each one of each of these two observers' minds will develop some particular *belief* about whether or not the outcome of the *other* observer's measurement was correlated or *anti*-correlated with the outcome of *her own*. And the *probability* of developing any particular such belief (for each of the two observers separately) is going to be precisely the usual quantum-mechanical one. And (as I said before) there simply isn't going to *be* any matter of fact about whether or not the outcomes of these two measurements, or the beliefs of these two observers, are ever "really" correlated with one another.

An Epistemological Remark

One of the things that the many-minds interpretation entails (as I've already mentioned) is that the beliefs of any sentient observer about the overall quantum state of the world will typically be mistaken.

Nothing, even in principle, can be done about that. No matter how much the observer in question knows of what the true laws of the world are, and no matter what observables she is capable of measuring, there can't be any experimental means whatever (as a little reflection will show) of reliably finding out what the overall quantum state of the world is, or (for that matter) what the quantum state of anything *in* the world is.

The sum total of what any such observer can conclude about the overall quantum state of the world (or, more precisely: the sum total of what any particular one of such an observer's *minds* can conclude about the overall quantum state of the world), from the outcomes of whatever experiments she does, is that that state (whatever it is) is not orthogonal (that is: not *perfectly* orthogonal) to the *effective* state that those outcomes *pick out*. And that's all.

And that's not much.

And that fact has curious consequences. It turns out, for example, that the Lorentz-covariance of the dynamical equations of motion of relativistic quantum field theories requires that the state that's associated with the *vacuum* in theories like that is necessarily not quite perfectly orthogonal to states in which there are electrons and baseballs and people and buildings (and all the other stuff we're used to) around.

And so what we've just been talking about is going to entail that on relativistic-field-theoretic versions of (say) a many-minds picture, nothing in our empirical experience (that is: nothing about the histories of our phenomenal states) is incompatible with the hypothesis that the quantum state of the universe is (for now and for all time) that *vacuum* state!

And that will throw an odd light (for example) on questions about where the universe initially *came* from.

But going through the details of all this would require a more technical discussion than I want to get into just now.[19]

19. I hate lines like that. But let the reader take note that this is the only one of them in this book.

Anyway, a slightly less incomplete account of this stuff (together with some further references and some remarks about how these ideas are and aren't related to the literature on the possibility that the universe is a vacuum fluctuation) can be found in a little paper of mine from a couple of years ago (Albert, 1988).

··· 7 ···

Bohm's Theory

There's an entirely different way of trying to understand all this stuff (a way of being absolutely deviant about it, a way of being polymorphously heretical against the standard way of thinking, a way of tearing quantum mechanics all the way down and replacing it with something else) which was first hinted at a long time ago by Louis de Broglie (1930), and which was first developed into a genuine mathematical theory back in the fifties by David Bohm (1952), and which has recently been put into a particularly clear and simple and powerful form by John Bell (1982), and that's what this chapter is going to be about.

Bohm's theory has more or less (but not exactly) the same empirical content as quantum mechanics does,[1] and it has much of the same mathematical formalism as quantum mechanics does too, but the metaphysics is different.

The metaphysics of this theory is exactly the same as the metaphysics of classical mechanics.

Here's what I mean:

This theory presumes (to begin with) that *every material particle in the world invariably has a perfectly determinate position.* And what this theory is about is the evolution of those positions in time. What this theory takes the *job* of physical science to be (to put it another way) is nothing other than to produce an *account* of those

---

1. Of course, there can't be *any* theory that has *exactly* the same empirical content as quantum mechanics does, since quantum mechanics (in its present unfinished condition, in the absence of any satisfactory postulate of collapse) doesn't *have* any exact empirical content!

evolutions; and the various *non*particulate sorts of physical things that come up in this theory (things like *force fields,* for example, and other sorts of things too, of which we'll speak presently) come up (just as they do in classical mechanics) only to the extent that we find we need to bring them up in order to produce the account of the *particle* motions.

And it turns out that the account which Bohm's theory gives of those motions is *completely deterministic.* And so, on Bohm's theory, the world can only appear to us to evolve *probabilistically* (and of course it *does* appear that way to us) in the event that we are somehow *ignorant* of its exact state. And so the very *idea* of probability will have to enter into this theory as some kind of an *epistemic* idea, just as it enters into classical statistical mechanics.

What the physical world consists of besides particles and besides force fields, on this theory, is (oddly) *wave functions.* That's what the theory requires in order to produce its account of the particle motions. The *quantum-mechanical wave functions* are conceived of in this theory as genuinely physical *things,* as something somewhat like force fields (but not quite), and anyway as something quite distinct from the particles; and the laws of the evolutions of these wave functions are stipulated to be precisely the linear quantum-mechanical equations of motion (always, period; wave functions never collapse on this theory); and the *job* of these wave functions in this theory is to sort of *push the particles around* (as force fields do), to *guide* them along their proper *courses;* and there are additional laws in the theory (new ones, *un*-quantum-mechanical ones) which stipulate precisely how they do that.[2]

---

2. Perhaps all this is worth spelling out in somewhat more pedantic detail. Here's the idea:

What *quantum mechanics* takes the wave function of a particle to be is merely a certain sort of mathematical representation of that particle's *state.*

What *this* theory takes the wave function of a particle to be, on the other hand, is a certain sort of genuinely physical *stuff.*

And the *physical properties* of such wave-functions-considered-as-stuff are (as with force fields) their *amplitudes* at every point in space.

And those amplitudes will invariably have determinate values (just as they invariably do, as purely mathematical objects, in quantum mechanics).

Setting Up

Let's start out by simply describing the mathematical formalism of the theory.

We'll begin with the case of a single structureless particle that's free to move around in only a single spatial dimension (since that case will turn out to be a particularly simple one, and a particularly instructive one), and then we'll build up from there.

The theory stipulates that the *velocity* of a particle like that at any particular time is given by the value of something called the *velocity function* $V(x)$ (which is a function of *position*, like the wave-function), evaluated at the point P at which the particle happens to be *located* at the time in question. Moreover, the theory stipulates that the velocity function for any particle at any time can be obtained (as a matter of law) from its *wave function,* at that *same* time, by means of a certain definite *algorithm*.

Let's set up a notation. Call the algorithm $V\{ \ldots \}$, where the symbol $\{ \ldots \}$ will serve as an empty *slot* into which any mathematical function of position can in principle be inserted.[3] What the theory is saying is that the velocity function $V(x)$ for any single structureless particle, moving around in a single spatial dimension,

---

It will frequently happen, of course (since the evolutions of wave functions here will invariably proceed in accordance with the dynamical equations of motion), that the wave function of a certain particle, at a certain instant, will fail to be an eigenfunction of one or another of the operators which the quantum-mechanical formalism associates with physical observables; but that will merely connote something about the *shape* of the wave-function-stuff here, something about its mathematical form, and *not* (as in quantum mechanics) that there is some *physical property of the world,* or some *potential* physical property of the world, which somehow *lacks a definite value!*

But of all this more later.

3. This algorithm, by the way, is stipulated to be precisely the same under all circumstances, irrespective of what the mass of the particle is, or what its charge is, or what forces it is subject to, or how those forces depend on space and time, or anything. All of those factors, of course, *do* play a role in determining the evolution of the *wave function* (since they all enter into the dynamical equations of motion); what they do *not* play any role in, however, is the way in which *velocity function* is obtained *from* the wave function.

at any particular time, will (as a matter of law) invariably be equal to the function $V\{\psi(x)\}$, where $\psi(x)$ is the wave function of that same particle at that particular time. And the theory further stipulates (as I mentioned above) that the *velocity* of that particle, at that time, will be equal to the number $V(P)$, where P is the *position* of the particle at that time.

And so if we're given the wave function of such a particle at some particular time t, and if we're also given its position at that time, then we can straightforwardly calculate its *velocity* at that time (and it's in this sense that the wave function can be said to push the particle around, to tell the particle where to go next).

And of course *that* will suffice, in principle, to calculate the position of the particle just an instant *after t* (since the velocity is just the rate of change of that position); and the *wave function* of the particle at that next instant can in principle be calculated too, with certainty, in the usual way, from the quantum-mechanical equations of motion; and so the *velocity* of the particle at that next instant can be obtained as well, and then the whole process can be repeated once again (in order to calculate the position of the particle and its wave function at the instant after *that*), and so on (as many more times as we like, up to whatever point in the future we choose).

And so the position of a particle and its wave function at any given time can in principle be calculated *with certainty*, on this theory, from its position and its wave function at any *other* time, given the external forces to which that particle is subject in the interval between those two times.

Suppose that the wave function of a particle at some particular initial time t is $\psi(x,t)$, and suppose that all of the forces are fixed in advance, and consider, in such circumstances (with the initial wave function held fixed, and with all of the external forces held fixed), how the motion of the particle will depend on its initial position.

Different initial positions will of course (in accordance with the laws of motion of this theory) pick out different *trajectories;* and each one of those different trajectories (in virtue of what it *means* to be a trajectory) will pick out some particular determinate position for the particle at every particular instant *later* than *t*.

Imagine a gigantic *swarm* of possible initial positions, and imagine the swarm of positions at some particular *later* time *t'*, which that *initial* swarm (in accordance with the laws of this theory, given the initial wave function and given the external forces) evolves *into*.

And now imagine a very particular *sort* of swarm of possible initial positions. Remember (from Chapter 2) that the conventional quantum-mechanical prescription stipulates that the probability that a measurement of the position of a particle will find the particle to be located at any particular point in space will be equal to the absolute value of the square of the particle's *wave function,* at the time of the measurement, at that point. Imagine, then, that our swarm of possible initial positions happens to be distributed (with respect to the wave function) just as the quantum-mechanical probabilities are: imagine that the *density in space* of the possible positions in the initial swarm happens to be everywhere equal to the square of the absolute value of the initial wave function.

If that's so (and this is the punch line), then it can be shown to follow from the dynamical equations of motion for the wave function and from the form of the algorithm for the calculation of the velocity functions that the density in space of the positions in the *later* swarm (no matter *what* later time *t'* happens to be) will likewise be everywhere equal to the square of the absolute value of the wave function *then* (at the *later* time). That's what's depicted in figure 7.1.

Here's another way to put it. Consider the following fairy tale. Suppose that the form of a certain single-particle wave function at a certain initial moment *t* is $\psi(x,t)$; and suppose that at just that particular moment God places the *particle* associated with that wave function (which was *absent* before, in this fairy tale) somewhere in the world, and suppose that God makes use of some genuinely *probabilistic* procedure for deciding precisely where to *put* that particle, and suppose that that procedure happens to entail that the probability that the particle gets put at any particular point is equal to the square of the absolute value of the particle's wave function at that point. And suppose that thereafter God does no

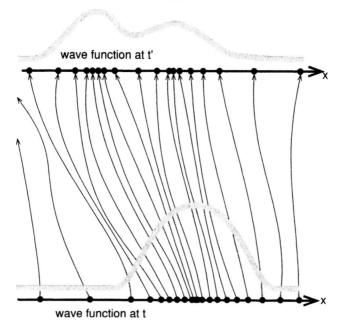

wave function at t'

wave function at t

Figure 7.1

more meddling and allows everything to evolve strictly in accordance with the deterministic laws of Bohm's theory. Then it will be the case that the probability that the particle will be located at any particular point in space at any *later* time (given the way things were initially set up) will be equal to the square of the absolute value of the particle's wave function at that point at that later time.

Here's a more compact way to put it: What happens (and this is what the algorithm was explicitly cooked up in order to *guarantee;* and, as a matter of fact, this is a way of uniquely specifying precisely what that algorithm *is*) is that *the particle gets carried along* with the *flows of the quantum-mechanical probability amplitudes in the wave function,* just like (say) a cork floating on a river.

Here's how to parlay that into something useful:

There is what amounts to a fundamental postulate of Bohm's

theory (let's call it the *statistical* postulate) which stipulates that if you're given the present wave function of a certain particle, and if you're given no information whatever about the present *position* of that particle, then, for the purposes of making calculations about motions of the particle in the future, what ought to be supposed, what ought to be plugged in to the formalism, is (precisely as it is in the fairy tale) that the *probability* that the particle is presently located at any particular point in space is equal to the square of the absolute value of its present *wave function* at that point.

That will guarantee (because of how the velocity algorithm works, because of how the particles always get carried along with the probability flows) that such calculations, under such circumstances, will necessarily entail that the probability that the particle will be located at any particular point in space at any particular *future* time will invariably be equal to the square of the absolute value of its wave function *then* (at the *future* time) at that point. And so in the event that we're given the present wave function of a single isolated particle, and we're given the external forces to which that particle is going to be subject, and we're given no information whatever (over and above what can be inferred, by means of the statistical postulate, from the wave function) about the present *position* of the particle (and it will turn out that that's more or less *all* of the information about the position of this sort of a particle that we can *ever* be in possession of, but of that more later), then Bohm's theory will entail precisely what principles C and D of Chapter 2 entail (that is: it will entail precisely what we know by experiment to be *true*) about the probability of *finding* that particle (if we should happen to *look* for it) at any particular point in space at any particular future time.

And so now we're apparently beginning to get somewhere.

The technical business of generalizing these laws, so as to be able to apply them to more complicated systems, goes (briefly) as follows:

(I) Three Space Dimensions. This part is trivial. The wave function (of a single isolated structureless particle, still) will now take on

values at every point in *three*-dimensional space (it will have the form $\psi(x,y,z)$), and the algorithm for calculating the velocity function will now become three algorithms (each of which looks much like the one-dimensional algorithm) for calculating three functions (one of which gives the velocity in the x direction, another of which gives the velocity in the y direction, and the third of which gives the velocity in the z direction), and each of those three functions will now be a function of three spatial positions. So the formalism will look like this:

(7.1) $V_x\{\psi(x,y,z)\} = V_x(x,y,z)$

$V_y\{\psi(x,y,z)\} = V_y(x,y,z)$

$V_z\{\psi(x,y,z)\} = V_z(x,y,z)$

The evolution here will be just as deterministic as it was in the one-dimensional case. And the structure of these algorithms will entail that the particle will now get carried along with the flows of the quantum-mechanical probability amplitudes in the wave function in three-dimensional space. And there will be a straightforward generalization of the statistical postulate to the three-dimensional case, with precisely the same sorts of consequences as were described above.

(II) Spin. This part is pretty simple too. Spin properties, to begin with, are taken here to be mathematical properties of the *wave functions*, and the idea is more or less that *those* properties, *those* parts of the wave function, play no direct role whatever in the determination of the velocity functions. The idea is that the velocity functions are to be determined here *precisely* as in (7.1), from the *coordinate-space* wave function of the particle *alone*.

That will entail, for example, that the velocity functions for a particle whose quantum-mechanical state vector is

(7.2) $|\text{black}\rangle|\psi(x,y,z)\rangle$ or $|\text{white}\rangle|\psi(x,y,z)\rangle$

or $|\text{whatever you like}\rangle|\psi(x,y,z)\rangle$

will be

(7.3) $V_i(x,y,z) = V_i\{\psi(x,y,z)\}$ $(i = x, y, z)$

The rule for handling states like

(7.4) $a|\text{black}\rangle|\psi(x,y,z)\rangle + b|\text{white}\rangle|J(x, y, z)\rangle$

in which (since (7.4) is *nonseparable* between its coordinate-space parts and its spin-space parts) there isn't any such thing as *the* coordinate-space wave function of the particle, is that *each* of its various *different* coordinate-space wave functions gets to *contribute* to determining the velocity-functions in proportion to what you might call its "weight"; i.e.,

(7.5) $$V_i(x,y,z) = \frac{|a\psi(x,y,z)|^2 V_i\{\psi(x,y,z)\} + |bJ(x,y,z)|^2 V_i\{J(x,y,z)\}}{|a\psi(x,y,z)|^2 + |bJ(x,y,z)|^2}$$

(III) Multiple-Particle Systems. This part is a little more complicated.

The first thing to talk about is the generalization of the conventional quantum-mechanical formalism of wave functions to cases of multiple-particle systems. That's pretty straightforward. The idea is just to write down something analogous to equation (2.27).

Suppose, then, that $|\psi_{1,2}\rangle$ represents an arbitrary quantum state of a two-particle system, and suppose that $|X_1 = x, X_2 = x'\rangle$ represents a state of that same system wherein particle 1 is localized at the (three-dimensional) point x, and wherein particle 2 is localized at the (three-dimensional) point x'. Then the two-particle *wave function* associated with the state $|\psi_{1,2}\rangle$ is defined to be

(7.6) $\psi(x,x') = \langle\psi_{1,2}|X_1 = x, X_2 = x'\rangle$

considered as a function of x and x'.

And just as anything whatever that can be said of the quantum states of single particles can be translated into the language of single-particle wave functions, anything whatever that can be said of the quantum states of two-particle systems can be translated into the language of two-particle wave functions.

Note, by the way, that whereas the single-particle wave functions are functions of position in a three-dimensional space, the *two*-particle wave functions can be looked at as functions of position in a somewhat more abstract *six*-dimensional space. The first three of those six dimensions will refer to the space of possible locations of particle 1, and the *second* three of those six dimensions will refer to the space of possible locations of particle 2; and so picking out any particular *point* in that six-dimensional space will amount to picking out particular values for the locations of *both* of those particles.

The laws of Bohm's theory for two-particle systems which stipulate precisely how the two-particle wave functions push such pairs of particles around are formulated just as if it were a *single* particle that were being pushed around in a *six*-dimensional space. There will now be six algorithms (each of which looks much like the one-dimensional algorithm) for calculating six velocity functions (one for the velocity in each of the three spatial dimensions for each of the two particles), and (this is important) each of those six functions will be a function of position in the six-dimensional space. So the formalism will look like this:

7.7) $\quad V_i(x,y,z,x',y',z') = V_i\{\psi(x,y,z,x',y',z')\} \quad (i = x,y,z,x',y',z')$

And the way to get the *velocities* out of the velocity *functions* is to plug in the position of the *two*-particle system in the *six*-dimensional space.

The calculations of the wave functions and the positions of two-particle systems at arbitrary times from those wave functions and positions at any particular initial time (given the external forces to which the particles are subject in the interval between those two times) will proceed, as we shall see, very much as they do for single-particle systems, and (of course) with just as much certainty as they do for single-particle systems. It will now be the position of the two-particle system in the six-dimensional space that gets carried along with the flows of the quantum-mechanical probabilities, in that same six-dimensional space, in the wave function.

And what the statistical postulate will dictate here is that in the event that all we initially know of a certain two-particle system is its wave function (and that will turn out, once again to be all that we can *ever* know of such systems), then what ought to be supposed (for the purpose of making calculations about the future positions of the particles) is that the probability that the two-particle system was initially located at any particular point in the six-dimensional space is equal to the square of the absolute value of the wave function of the system, at that initial time, at that same particular point in the six-dimensional space (and note that this will reduce precisely to the *one*-particle statistical postulate in the event that the wave function of the two particles happens to be *separable* between them).

And that will guarantee (because of how these systems always get carried along with the probability flows) that such calculations, under such circumstances, will necessarily entail that the probability that the system will be located at any particular point in the six-dimensional space at any particular *future* time will invariably be equal to the square of the absolute value of its wave function *then* (at the *future* time) at that point. It will guarantee (to put it slightly differently) that such calculations, under such circumstances, will necessarily entail precisely the same things as the two-particle versions of principles C and D of Chapter 2 entail about the probability of *finding* the two particles (if we should happen to look for them) at any particular pair of points in the ordinary *three*-dimensional space at any particular future time.

And the reader will now have no trouble in constructing higher-dimensional formalisms, along precisely the same lines, for treating systems consisting of arbitrary numbers of particles, and even (in principle) for treating the universe as a whole.

And the statistical postulate, in a formalism like *that,* can be construed as stipulating something about the *initial conditions* of the universe; it can be construed (in the fairy-tale language, say) as stipulating that what God did when the universe was created was first to choose a wave function for it and sprinkle all of the particles into space in accordance with the quantum-mechanical probabilities, and then to leave everything alone, forever after, to evolve deterministically. And note that just as the one-particle postulate can be derived (as we saw above) from the *two*-particle postulate, *all* of the more specialized statistical postulates will turn out to be similarly derivable from *this* one.

The Kinds of Stories the Theory Tells

Let's start slow.

Let's look (to begin with) at some measurements with spin boxes.

Consider, for example, an electron, whose wave function is black; it is situated in coordinate space as pictured in figure 7.2, and it is (as shown in that figure) on its way into a *hardness* box.

Given all that, all of the *future* positions of this electron, on this theory, can in principle be determined, with certainty, from its *present* position (and so the *aperture* through which this electron will ultimately *exit* this box can in principle be determined from its present position, and so the *outcome* of the upcoming measurement

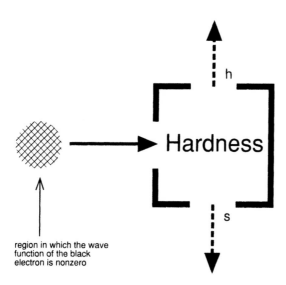

region in which the wave
function of the black
electron is nonzero

Figure 7.2

of the *hardness* of this electron can in principle be determined from its present position).

Let's see how that works. Suppose that the configurations of the hardness-dependent forces inside the hardness box are such as to cause *eigenfunctions* of the hardness to evolve as depicted in figure 7.3. Then the evolution of the wave function in question *here,* the *black* wave function, will of course be a linear *superposition* (with a plus sign and with equal coefficients) of the two evolutions

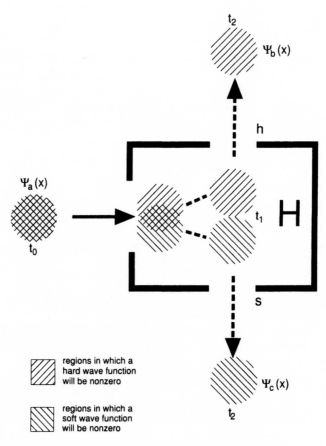

Figure 7.3

depicted there; so that (under these circumstances, in the time it takes for the wave function to pass all the way through the box):

$$(7.8) \qquad |\text{black}\rangle|\psi_a(x)\rangle \rightarrow \frac{1}{\sqrt{2}}(|\text{hard}\rangle|\psi_b(x)\rangle + |\text{soft}\rangle|\psi_c(x)\rangle)$$

where $\psi_a(x)$ is a wave function which is nonzero only in the vicinity of the point a and which is headed to the right; $\psi_b(x)$ is a wave function which is nonzero only in the vicinity of the point b and which is headed *up*; and $\psi_c(x)$ is a wave function which is nonzero only in the vicinity of the point c and which is headed *down*.

Now consider how the motion of the electron is going to depend on its initial position. And keep in mind that what happens on this theory is that the electron, wherever it happens to be, invariably gets carried along with the local currents of the quantum-mechanical probability amplitudes. And note that the situation here has been cooked up so as to be perfectly symmetrical between the hard and the soft branches of the wave function. And so (for as long as the hard and the soft branches *overlap*, which will be up to time t_1 in figure 7.3) whenever the electron is located in the *intersection* of the two branches, then its velocity in the vertical direction will patently be *zero;* and whenever the electron is located exclusively in *one* or the *other* of those two branches (whenever, that is, the electron is located in a region of space in which the value of one of these two wave functions is *zero*), then it will invariably move perfectly along with that particular branch (the one that *isn't* zero there).[4] And all of that (if you think about it) will entail that in the event that the electron starts out in the *upper* half of the region where $\psi_a(x)$ is nonzero, then it will ultimately emerge from the *hard* aperture of the box; and in the event that the electron starts out in the *lower* half of that initial region, then it will ultimately emerge from the *soft* aperture of the box. It's as simple as that.

And note that in the event that the electron emerges from the hardness box through the *hard* aperture and is subsequently fed

4. The statistical postulate will guarantee that there will be no probability whatever of the electron's ever being located in a region of space in which the values of *both* of those wave functions will be zero.

into *another* hardness box (as in figure 7.4), then it will with certainty emerge from that *second* box through the hard aperture as well; and in the event that the electron emerges from the hardness box through the *soft* aperture and is subsequently fed into another hardness box (as in figure 7.5), *then* it will with certainty emerge from that second box through the *soft* aperture. Here's why: Once the electron gets out of the first box, if it emerges, say, through the

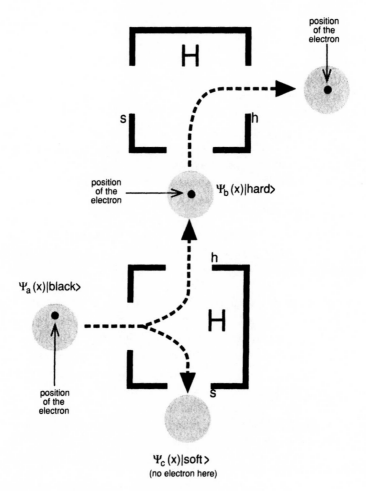

Figure 7.4

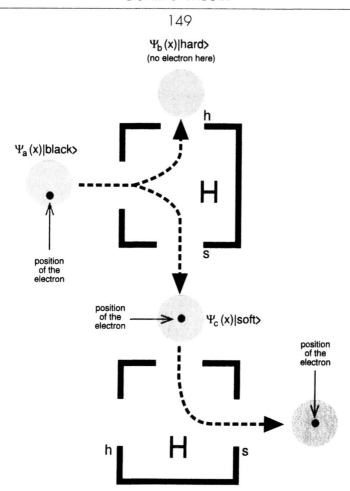

$\Psi_b(x)|hard\rangle$
(no electron here)

$\Psi_a(x)|black\rangle$

position
of the
electron

position
of the
electron

$\Psi_c(x)|soft\rangle$

position
of the
electron

Figure 7.5

hard aperture, then (unless the hard and soft branches of the wave function are *reunited* in space *later on,* by means of reflecting walls, perhaps; but we'll talk about that in a minute) it will subsequently be moving in regions of space in which the amplitude of the *soft* part of its wave function is *zero;* and so that soft part will have *no effect whatever* on its subsequent motions; and it will simply be carried along through the second hardness box (and through whatever else it may encounter thereafter) with the *hard* part of its wave

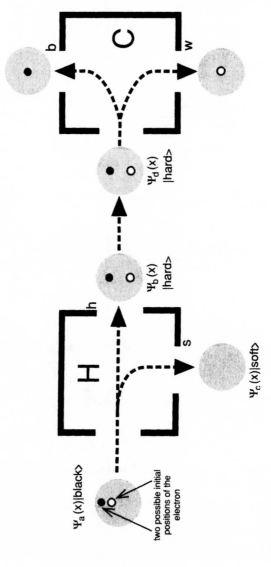

Figure 7.6

function. And of course there is an analogous story to tell in the event that the electron emerges through the *soft* aperture of the first box.

And the reader will now be able to confirm for herself that in the event that the electron emerges from the hardness box (the *first* one) through, say, the hard aperture and is subsequently fed into a *color* box (as shown in figure 7.6), *then*, in the event that the electron is located in the *upper* half of the region in which $\psi_d(x)$ is nonzero, it will ultimately emerge from the *color* box through the *black* aperture, and in the event that it is located in the *lower* half of that region, it will ultimately emerge from the color box through the *white* aperture.

And in the event that the two branches of the wave function emerging from the two different apertures of the hardness box are ever *reunited*, by means, say, of an arrangement of reflecting walls and a "black box" like the one in figure 7.7, if the electron is *then* fed into a color box, *then*, no matter *what* position it initially had in the region in which $\psi_a(x)$ was nonzero, it will with certainty emerge from that color box through its *black* aperture. And if a wall is inserted somewhere along one of the two paths, then certain initial positions of the electron within $\psi_a(x)$ will entail that the electron never reaches the color box *at all*, and certain *other* such positions will entail that the electron finally emerges from the *white* aperture of the color box, and certain *other* such positions will entail that the electron will finally emerge from the *black* aperture of the color box.

And in the event that all that we initially happen to *know* of this electron is its *wave function* (which, again, will turn out to be all we *can* know of it, but of that more later), then the statistical postulate will straightforwardly reproduce all of the familiar quantum-mechanical frequencies of the various different possible outcomes of experiments like these (since the outcomes of experiments like these invariably come down to facts about the final *positions* of the *measured particles*). And that's going to turn out to be an instance of something a good deal more general: The fact that Bohm's theory gets everything right about the *positions* of things is going to entail that it *also* gets everything right about the out-

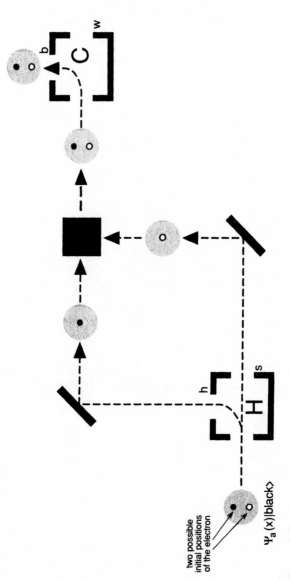

Figure 7.7

comes of any measurements of any quantum-mechanical observables whatever, so long as those outcomes get *recorded* (at one point or another) in the positions of things;[5] but a good deal more will need to be said, later on, about that too.

Here's something curious: What happens (as we saw before) when things are set up as in figure 7.2, and if the wave function of the electron is initially *black* and if the initial position of the electron is, say, within the *upper* half of the region where $\psi_a(x)$ is nonzero, is that the electron ultimately emerges from the *hard* aperture of the box.

Consider how all this depends on the *orientation* of the *hardness box*. Suppose that everything starts out just as described above (the wave function of the electron is black, the position of the electron is in the upper half of $\psi_a(x)$), except that the *hardness box* is *flipped over*, as in figure 7.8. Then the evolution of the wave function, as it passes through the box, will proceed as follows:

(7.9) $|black\rangle|\psi_a(x)\rangle \rightarrow \frac{1}{\sqrt{2}}(|soft\rangle|\psi_b(x)\rangle + |hard\rangle|\psi_c(x)\rangle)$

---

5. And note (by the way) that that's a good deal *more* than can be said for the GRW theory (which is to say that it's a good deal more than can be said for *any* existing theory of the collapse).

The experiments that the GRW theory gets everything right about (after all) are just the ones whose outcomes get recorded in the position of something *macroscopic,* but the experiments that *Bohm's* theory gets everything right about are the ones whose outcomes get recorded in the position of *anything at all* (macroscopic or *microscopic*).

Bohm's theory, for example, is going to turn out to make the right predictions about the outcomes of experiments like the one depicted in figure 5.4, since the outcome of that sort of an experiment gets recorded in the *position* of the measured electron (the GRW theory, remember, turned out to entail that experiments like that, at least in their pre-retinal stages, don't have any outcomes at all); and Bohm's theory is also going to turn out to make the right predictions about the outcomes of experiments like the one depicted in figure 5.7, since the outcome of *that* sort of an experiment ultimately gets recorded in the position of the particle in the middle of John's head (and remember that *these* sorts of experiments turned out to be problematic for collapse theories *in general*: it turned out that there can't be *any* theory of the collapse on which *this* sort of an experiment has an outcome).

Nonetheless, it remains to be seen (to say the least) whether or not Bohm's theory makes the right predictions about the outcomes of absolutely all experiments whatsoever; but of that more later.

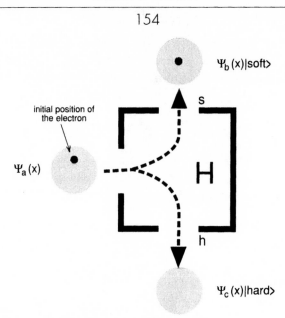

$\Psi_b(x)|soft\rangle$

initial position of
the electron

$\Psi_a(x)$

s

H

h

$\Psi_c(x)|hard\rangle$

Figure 7.8

and the path of the particle through space (if you think it through) will patently be precisely the *same* as it would have been if the box *hadn't* been flipped, and so in *this* case the electron is ultimately going to leave the box through the *soft* aperture. And in the event that the electron starts out in the *lower* half of $\psi_a(x)$, with everything else as above (with the box flipped), *then* the electron will ultimately emerge through the *hard* aperture.

And so (even though this is a completely deterministic theory) the outcome of this sort of a "measurement" of the hardness of an electron will in general *not* be pinned down in this theory even by means of a complete specification of the electron's position and its wave function, which is (after all) *everything there is* to be specified about that electron. The outcome of such a hardness measurement is in general going to depend, on this theory, on precisely *how* and under precisely what *circumstances* the hardness gets measured, even down to the orientation of the hardness box in space.

And so it doesn't quite make sense, in general, on this theory, to think of hardness as an intrinsic property of electrons or of their wave functions (or of any combination of the two) at all. Properties like that (properties which, for these sorts of reasons, can't quite be thought of as intrinsic to the systems in question) have come to be referred to in the literature (for obvious reasons) as *contextual*.

And it turns out that color and gleb and scrad and momentum and energy and *every one* of the traditional quantum-mechanical observables of particles other than *position* are contextual properties on this theory too. As a matter of fact, there are theorems in the literature to the effect that *any* deterministic replacement for quantum mechanics whatever will invariably have to treat certain such observables as contextual ones.[6]

Note, however, that in the event that the wave function of an electron happens to be an *eigenfunction* of the hardness, then the outcome of a hardness measurement carried out on that electron will patently *not* depend on the orientation of the hardness box, and not on any *other* particulars of the condition of the hardness measuring device either, so long as whatever device that *is* satisfies the requirements for being a "good" device for measuring the hardness.

Similarly, if an electron is fed through one hardness box and then directly through *another,* it will with certainty emerge from both of those boxes through the *same* aperture, *irrespective* of the orientations of those two boxes or of anything else about them (so long as they're both *hardness boxes*), unless the hard and the soft branches of the electron's wave function have had a chance to overlap *in between* the two boxes.

And of course the same sorts of things are true of color as well, and analogous things are true of every other quantum-mechanical observable too.

For systems consisting of more than a single particle, Bohm's laws become explicitly *nonlocal.*

6. See, for example, Gleason, 1957, and Kochen and Specker, 1967.

Here's how that happens: Consider a two-particle system. The idea is that *each* of the *six* velocity functions for a two-particle system is going to depend, in general (as I mentioned before), on the location of the *entire* two-particle system in the *six-dimensional space*. And so (for example), in a system consisting of particle 1 and particle 2, once the wave function is fixed, the velocity in the x-direction of particle 1, at some particular moment, is in general going to depend not only on the position of particle 1 at that moment but *also* on the position of particle 2, at that *same* moment, no matter how far away particle 2 may (at that moment) happen to be! And so the *motions* of particle 2 (which will have the effect of changing the location, in the *six*-dimensional space, of the entire *two*-particle system) will in general play a very direct role, *instantaneously* (no matter how far apart the two particles may happen to be, or what may lie between them), in determining the velocities of particle 1. And (of course) vice versa.

Now, it turns out that in the event that the two-particle wave function is *separable* between the two particles, then these sorts of nonlocalities won't arise. In the event that the wave function is separable between the two particles, then it turns out that (once the wave function is given) the velocity of particle 1 always depends only on the position of particle 1 and the velocity of particle 2 depends only on the position of particle 2, and (more generally) the two-particle theory reduces completely, in that event (and if the particles don't interact with one another by means of ordinary *force fields*), to a pair of one-particle theories, just as in ordinary quantum mechanics. And if *that* weren't so (come to think of it), the *one*-particle theory wouldn't make any sense at all!

But things get more interesting if the wave function of the two particles is *non*separable.

Consider, for example, a pair of electrons; and suppose that the coordinate-space part of the wave function of one of those electrons (electron 1) is initially nonzero only in the vicinity of point a, and suppose that the coordinate-space part of the wave function of the *other* electron (electron 2) is initially nonzero only in the vicinity of point f (see figure 7.9); and suppose that the *spin-space* part of

$\Psi^1_a(x)$ $\Psi^2_f(x)$

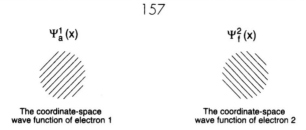

The coordinate-space   The coordinate-space
wave function of electron 1   wave function of electron 2

Figure 7.9

the wave function of this system is of the (nonseparable) EPR type, so that the wave function *as a whole* looks like this:

(7.10) $\frac{1}{\sqrt{2}}(|\text{hard}\rangle_1|\psi_a(x)\rangle_1|\text{soft}\rangle_2|\psi_f(x)\rangle_2 - |\text{soft}\rangle_1|\psi_a(x)\rangle_1|\text{hard}\rangle_2|\psi_f(x)\rangle_2)$

And suppose that electron 1 happens to start out in the upper half of the region in which $\psi_a(x)$ is nonzero and that it is now passed through a right-side-up hardness box, as in figure 7.10, so that the wave function of the two-particle system becomes:

(7.11) $\frac{1}{\sqrt{2}}(|\text{hard}\rangle_1|\psi_b(x)\rangle_1|\text{soft}\rangle_2|\psi_f(x)\rangle_2 - |\text{soft}\rangle_1|\psi_c(x)\rangle_1|\text{hard}\rangle_2|\psi_f(x)\rangle_2)$

Electron 1 (for precisely the same reasons as in the single-particle case) will end up in the vicinity of point *b*. And once that's the case (and this is the punch line), the two-electron system will be located at a point in the six-dimensional space at which the value of the second term in (7.11) is 0. And so, thereafter (unless, or until, the two branches of the wave function subsequently come again to overlap in the six-dimensional coordinate space), that second branch will have *no effect whatever* on the motions of *either* of these electrons: electron 1 will behave, under all circumstances, as if its wave function were purely hard, and *electron 2 will behave, under all circumstances, as if its wave function were purely soft.* Electron 2, for example, will now emerge from any hardness box it gets fed through by the *soft* (as in figure 7.11) aperture, no matter *what* the orientation of that box or the initial position of that

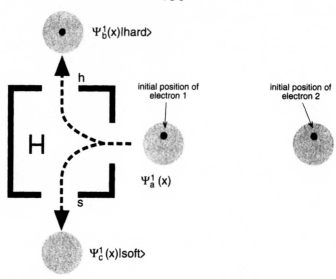

Figure 7.10

electron may happen to be. And so, even though the wave function evolves here entirely in accordance with the linear dynamical equations of motion, the passage of electron 1 through the hardness box brings about (as it were) an *effective collapse* of the wave function of the entire two-electron system, *instantaneously,* no matter how far apart they may happen to be or what may happen to lie between them.

And in the event that all we initially know of electrons like this one is that their wave function is the one in equation (7.10), then (as the reader can now easily confirm for herself) the statistical postulate and the Bohm-theoretic equations of motion will straightforwardly reproduce all of the standard quantum-mechanical predictions about the outcomes of measurements with spin boxes on EPR systems. And of course *that* entails (by means of Bell's theorem) that *some* sort of nonlocality was going to *have* to come up in this theory. But note that the particular sort of nonlocality that *does* come up in this theory turns out (compared, say, to the nonlocality in *quantum mechanics*) to be quite astonishingly *con-*

crete; it turns out (that is) to be a genuinely physical sort of *action at a distance.*

Suppose, for example, that electron 1 in figure 7.10 starts out (as above) in the upper half of the region in which $\psi_a(x)$ is nonzero, and suppose that electron 2 starts out in the upper half of the region in which $\psi_f(x)$ is nonzero. If that's so, and if electron 1 gets sucked through a right-side-up hardness box, as in figure 7.10, then (as we've just seen) electron 1 will end up effectively hard, and electron 2 (whatever *its* initial position is) will instantaneously become effectively *soft.* But of course in the event that electron 2 gets sucked through a hardness box first, and if *that* hardness box happens to be right-side-up, then (since electron 2 is in the upper half of $\psi_f(x)$) electron 2 will end up effectively hard, and electron 1 will instantaneously become effectively *soft* (and so if electron 1 *subsequently* gets sucked through a hardness box, with *any* orientation, then *it* will with certainty emerge by the *soft* aperture).

And if everything is initially as I just described, and electron 1

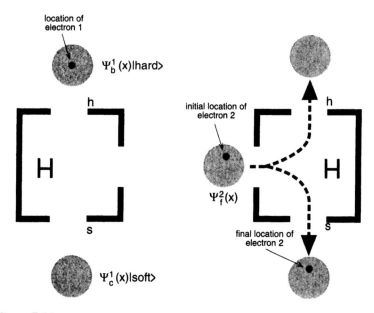

Figure 7.11

gets sucked through an *upside-down* hardness box, *then* electron 1 will end up effectively *soft* and electron 2 will instantaneously become effectively *hard*.

And so, if we're given the (nonseparable) wave function of such a system, and if we're given the positions of both of its constituent particles, then all this will patently afford a means of transmitting *discernible information,* instantaneously, over any distance, no matter what may lie in the way.

And so there can't possibly be any such thing as a Lorentz-co-variant relativistic extension of this sort of a theory. This sort of theory will invariably require a *preferred frame;* this sort of theory (to put it another way) will invariably require an *absolute standard of simultaneity.*

But of course in the event that all we know of such a system is its *wave function,* then (as I mentioned above) all of the familiar quantum-mechanical probabilities for the outcomes of experiments with spin boxes on that sort of system will reemerge, and the sort of nonlocal information transmission just described will become impossible, and it will become impossible to determine (by means of experiments on a system like that, in any *non*covariant relativistic extension of Bohm's theory) *which* Lorentz-frame is the *preferred* one.[7]

The statistical predictions of any relativistic extension of Bohm's theory, then (supposing that the universe initially got started in the fairy-tale way), will turn out to be fully Lorentz-covariant, even though the underlying theory *won't* be;[8] and so taking Bohm's

7. The idea (once again) is that the outcomes of these sorts of measurements are recorded in the final *positions* of the two *particles;* and we know that in the event that all the information we initially have about such a pair of particles is their *wave function,* then the Bohm-theoretic probability distributions for those positions are going to be precisely the same as the quantum-mechanical probability distributions for those positions.

8. Of course, none of this talk will matter much unless it turns out that some relativistic extension of Bohm's theory (that is: some Bohm-type replacement for relativistic quantum field theory) can actually be cooked up.

It isn't clear whether or not that can be done. It turns out to be hard (for example) to figure out what sorts of field variables can possibly stand in for the *positions of particles* in a theory like that; it turns out to be hard (that is) to figure out what

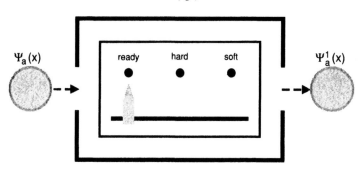

Figure 7.12

theory *seriously* will entail being *instrumentalist* about special relativity.[9]

Consider what happens on this theory in the event that the outcome of a measurement ultimately gets recorded in something macroscopic.

Consider, for example, the case of a hardness measuring device equipped with a macroscopic pointer, like the device depicted in figure 7.12. We've dealt with that sort of thing before.

---

sorts of field variables ought to get picked out by a theory like that as the *noncontextual* ones.

Bohm and Bell have both thought a good deal about these matters, but what they've come up with so far is only somewhat encouraging. Bohm (1952) has techniques for cooking up thoroughly Bohm-type replacements for a few particular field theories (but there are other such theories that those techniques *won't* work for); and Bell (1984) has a method which is a good deal more generally applicable but which generates theories that aren't completely *deterministic*.

Here's what the fundamental trouble is: Bohm's theory (as it presently stands) is quite deeply bound up with a very particular sort of *ontology*; the trouble (that is) is that this theory isn't a replacement for quantum theory *in general* (like the many-minds interpretation is), but only for those quantum theories which happen to be theories of *persistent particles;* and so the business of cooking up Bohm-type replacements for quantum theories of *other* sorts of systems *(field* systems or *string* systems or what have you) always has to proceed, without any guarantee of eventual success, case by case, from the ground up.

9. Bell (1976) began to explore what it's like to do that in a nice article called "How to Teach Special Relativity."

Suppose that a black electron is fed into that device, when the device is in its ready state. The wave function will evolve like this:

$$(7.12) \qquad |\psi_r\rangle_m(|\psi_a\rangle_e|\text{black}\rangle_e) \rightarrow$$

$$\tfrac{1}{\sqrt{2}}(|\psi_b\rangle_m(|\psi_b\rangle_e|\text{hard}\rangle_e) + |\psi_s\rangle_m(|\psi_b\rangle_e|\text{soft}\rangle_e))$$

where $|\psi_r\rangle_m$ is a state of the billions of particles in the pointer in which all of those particles are sitting more or less directly underneath the word "ready" on the dial (and similarly for $|\psi_b\rangle_m$ and $|\psi_s\rangle_m$), and $|\psi_a\rangle_e$ and $|\psi_b\rangle_e$ are the coordinate-space states of the electron depicted in figure 7.12.[10]

Once the electron passes all the way through the box and the wave function is the one on the right-hand side of equation (7.12), and the particles in the pointer are either under the "hard" on the dial or under the "soft" on the dial,[11] then the wave function of this composite system will have been effectively *collapsed* onto one or the other of the two terms on the right-hand side of equation (7.12), and (unless or until the two branches of the coordinate-space wave function of the particles in the pointer somehow *drift back together*) the electron will subsequently have an effectively determinate hardness.

And suppose that the two branches of the coordinate-space wave function of the particles in the pointer *do* eventually drift back together but that there happens to be an *air molecule* in the vicinity of the dial (as in figure 5.1), and that the *position* of that air molecule ends up *correlated* to the *hardness* of the electron (as in equation (5.4)). *Then* (unless or until the two branches of the coordinate-space wave function of the *air molecule* somehow drift

10. Note, to begin with, that the outcome of *this* sort of a hardness measurement is going to depend on the precise initial positions of the particles in the *pointer;* it's the *pointer* (and *not* the measured particle itself, as in figure 7.3) whose coordinate-space wave function gets split here; it's the pointer (and not the measured particle) whose position gets correlated to the hardness.

11. And note that either *all* of those particles will be under the "soft" or all of them will be under the "hard"; the form of the wave function (together with the statistical postulate) will entail that the probability that *some* of those particles end up in *one* of those places and some of them end up in the *other* will be *zero.*

back together) the electron will *still* have an effectively determinate hardness.

And even if the various branches of the *air molecule's* wave function were to drift back together *too,* but records of the outcome of the hardness measurements were still to survive, say, in the positions of ink molecules on the pages of a lab notebook, or in the positions of a few ions in a few neurons in some experimenter's brain, or even in so much as the position a single subatomic particle anywhere in the universe at all (as in the EPR example), *then* (unless or until we can manage to arrange that all that altogether ceases to be true) the electron will *still* have an effectively determinate hardness.

And so the business of *undoing* the effective determinateness of the hardness of this electron, or of empirically confirming that as a matter of fact that determinateness *is* merely effective, and that nothing has actually *collapsed,* will be (to say the least) quite fantastically difficult.[12]

And so Bohm's theory is going to make things look (for all practical purposes) as if wave functions *do* collapse when we do measurements with instruments with macroscopic pointers; and as a matter of fact, Bohm's theory is going to make things look as if wave functions collapse whenever we do measurements with any instruments whatever which (by one means or another) leave *records* of the *outcomes* of those measurements in the *positions of particles* in their *environments;*[13] and those collapses are going to appear to occur in more or less precisely the way we're *used* to, the

12. The difficulty here is of course precisely the same as the one we encountered on pages 88–92. The trouble is that the *environment* of a pointer like the one we're talking about here will act as a gigantic collection of extremely effective measuring devices for the pointer's *position;* and the business of confirming that as a matter of fact the Bohm wave function *hasn't* collapsed will involve either avoiding or reversing or somehow taking account of *every last one* of those measurements.

13. Note that these recordings need not be in the positions of anything *macroscopic;* what's important is merely that those recordings be in the positions of *something* and that they be (practically speaking) difficult to undo, or difficult to take account of. Consider, for example, the case of the measurement depicted in figure 5.4.

way we tried to *force* them to do in Chapter 5, even though it can *in principle* be empirically confirmed (on this theory) that in fact they don't occur at all.

And in the event that all we know of the systems involved are their *wave functions* at the moment just before the measurement interaction takes place, or in the event that all we know of those systems are their *effective* wave functions at the moment just before the measurement interaction takes place, then the statistical postulate will straightforwardly entail that the epistemic probabilities of the various possible effective collapses which that interaction may *bring about* will all be precisely the same as the *ontic* probabilities of the corresponding *actual* collapses (that is: the probabilities of principle D of Chapter 2) in *quantum mechanics*.

And it will turn out (as I've already mentioned) that this theory entails that all that we ever *can* know of the present states of such systems are their wave functions (or perhaps their *effective* wave functions); and *that's* how it happens (and I've already mentioned this too) that this theory has more or less the same empirical content as quantum mechanics does; that's how it happens (more particularly) that this theory entails that (even though the fundamental laws of the world are absolutely deterministic) we can never put ourselves in a position to predict any more about the outcomes of future experiments than the conventional quantum-mechanical *uncertainty relations* allow us to.

Let's see (finally) how that works.

Let's think through a simple example. Consider (say) an electron whose wave function happens to be $|\psi_a(x)\rangle|black\rangle$ and which is on its way into a hardness box, like the one in figure 7.2. Suppose that we initially know nothing of this electron other than its wave function; so that all that we're initially in a position to predict, with certainty, about the outcomes of spin measurements on this electron, is that (just as in quantum mechanics) any upcoming measurement of the electron's color will necessarily find it to be black.

Consider how things would stand if we were now somehow able to find out the *position* of this electron, or if we were merely able to find out, say, whether this electron is in the upper or in the lower half of the region where $\psi_a(x)$ is nonzero, without (in the course of finding that out) *changing* the electron's wave function. Then, of course, we would still be in a position to predict the outcome of any future measurement of the electron's color (since the wave function of the electron would still be the same), but we would *also* now be in a position to predict, with certainty, the aperture through which the electron would ultimately emerge from, say, a right-side-up *hardness* box; and so we would be in a position to violate the quantum-mechanical uncertainty relations; we would be in a position to predict *more* about this electron than quantum mechanics *allows* us to; we would be in a position to predict more about it than (as a matter of empirical fact) we find we *can* predict about it.

Let's figure out what's going on. Think, to begin with, about what it will involve (think, that is, about what sorts of *physical acts* it will involve) to find out whether the electron in the above scenario is located in the upper half or the lower half of the region in which $\psi_a(x)$ is nonzero. What we shall need to do is to bring some physical property of a measuring device (the position of its pointer, say) into the appropriate sort of *correlation* with the location of the electron. And what will need to be done in order to accomplish *that* (as a little reflection on the laws of this theory will show) is to bring about the analogous sorts of correlations in the *wave functions* of those two systems; what will need to be done (to put it another way) is to bring into being a wave function of the *composite* system which is *zero* in all those regions of the many-dimensional coordinate-space in which those correlations *fail* to obtain.

And so finding out whether the electron in the above scenario is located in the upper or the lower half of the region in which $\psi_a(x)$ is nonzero, *without* (in the process of finding that out) *changing the wave function,* is going to be (as a matter of *fundamental principle*) completely out of the question.

Suppose, for example, that we were to measure whether the electron is in fact in the upper or the lower half of the region in which $\psi_a(x)$ is nonzero (let's call that "region a" from here on), using a measuring device (m) like the one depicted in figure 7.13. Then (in accordance with the dynamical equations of motion) the

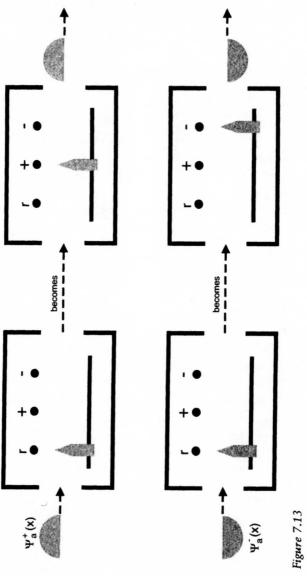

Figure 7.13

wave functions of the electron and the measuring device will evolve as follows:

(7.13) $|r\rangle_m|\psi_a(x)\rangle_e|\text{black}\rangle_e \rightarrow$

$\qquad |+\rangle_m|\psi_a^+(x)\rangle_e|\text{black}\rangle_e + |-\rangle_m|\psi_a^-(x)\rangle_e|\text{black}\rangle_e$

where $\psi_a^+(x)$ is equal to $\psi_a(x)$ in the upper half of region a and is 0 elsewhere, and $\psi_a^-(x)$ is equal to $\psi_a(x)$ in the *lower* half of region a and is 0 elsewhere, and of course $|r\rangle$ and $|+\rangle$ and $|-\rangle$ are the states of the *pointer* in which it's pointing at "r" or "+" or "−" on the dial, respectively.

And so in the event that the electron is located in the *upper* half of region a, then the pointer will with certainty end up pointing at "+" and the coordinate-space *wave function* of the electron will be effectively *collapsed* onto $\psi_a^+(x)$, and the statistical postulate will entail (by means of a straightforward conditionalization) that the probability that the electron is located at any particular point in space is equal to $|\psi_a^+(x)|^2$; and in the event that the electron is located in the *lower* half of region a, *then* the pointer will with certainty end up pointing at "−" and the coordinate-space wave function of the electron will be effectively collapsed onto $\psi_a^-(x)$, and the statistical postulate will entail that the probability that the electron is located at any particular point in space will be equal to $|\psi_a^-(x)|^2$.

And all of that will ineluctably *change* the way in which the *future motions* of this electron depend on its present position. In the event, for example, that the electron's coordinate-space wave function gets effectively collapsed onto $\psi_a^+(x)$, then (as the reader can easily confirm) the outcome of an upcoming measurement with a hardness box will depend on whether the electron is located in the top *quarter* (not half!) or the next-to-the-top quarter of the region a. Of course the outcome of our position measurement will give us no idea whatever which of *those* two is the case (that is: the epistemic probabilities will be precisely fifty-fifty); and so everything perversely conspires together here so as to insure that that position measurement will have done us precisely *no good at all,* in so far as the violation of the uncertainty relations is concerned. And of course the same sort of thing happens in the event that the electron's wave function gets effectively collapsed onto $\psi_a^-(x)$.

And as a matter of fact, this sort of thing turns out to constitute an absolutely general *law*: it turns out (and the reader will now be in a position to construct an argument for this for herself) that this theory entails that if we initially know nothing of a system (*any* system) other than its wave function (if, that is, we know nothing more of the system's *location* in its many-dimensional coordinate space than can be *inferred* from its wave function by means of the *statistical postulate*), then all we shall possibly be in a position to know at any particular future time of that system's location in that space (by means of measurements, for example, or by any physical means whatever) will be what follows (by means of the statistical postulate) from that system's *effective wave function* at that future time.

And if we apply all that to the universe as a whole, and if we presume that the universe was initially created in accordance with the fairy tale described earlier, then this theory will entail that all that can ever be found out by any observer whatever about the present Bohm-state of any particular physical system can always be completely *summed up* (with the help of the statistical postulate) in a *wave function* (sometimes this will turn out to be the *actual* wave function of the system in question, but more often it will merely be an *effective* one). And so the Bohm theory (even though it's a completely deterministic theory) will systematically and invariably and unavoidably *prohibit* us from ever predicting the outcomes of future measurements of the positions of particles any more accurately than those *wave functions* allow us to do; it will (that is) prohibit us from ever predicting those outcomes any more accurately than the *uncertainty relations* allow us to do.

That's how this theory manages to clean up after itself: that's what entails (for example) that we can't ever empirically discover that the outcomes of measurements with spin boxes depend on the *orientations* of those boxes (even though they *do* depend on those orientations, if this theory is right); and that's what entails that we can't ever empirically discover that the outcomes of measurements with spin boxes can *also* depend, *nonlocally* (when states like (7.10) obtain), on the orientations of fantastically *distant* spin boxes (even though they *can* depend on the orientations of distant boxes, if this theory is right); and that's what entails that we can't ever find out what the natural standard of absolute *simultaneity* is (even though there *is* a natural standard of absolute simultaneity, if this theory is right); and so on.

And so if this theory is right (and this is one of the things about it that's cheap and unbeautiful, and that I like), then the fundamental laws of the world are cooked up in such a way as to systematically *mislead* us about themselves.

Here's what's so cool about this theory:

This is the kind of theory whereby you can tell an absolutely low-brow story about the world, the kind of story (that is) that's about *the motions of material bodies,* the kind of story that contains nothing cryptic and nothing metaphysically novel and nothing ambiguous and nothing inexplicit and nothing evasive and nothing unintelligible and nothing inexact and nothing subtle and in which no questions ever fail to make sense and in which no questions ever fail to have answers and in which no two physical properties of anything are ever "incompatible" with one another and in which the whole universe always evolves *deterministically* and which recounts the unfolding of a perverse and gigantic conspiracy to make the world *appear* to be *quantum-mechanical.*

And that conspiracy works (in brief) like this: Bohm's theory entails everything that quantum mechanics entails (that is: everything that principles C and D of Chapter 2 entail, everything that we empirically know to be *true*) about the outcomes of measurements of the positions of particles in isolated microscopic physical systems; and moreover it entails that whenever we carry out a measurement of *any* quantum-mechanical observable *whatever,* then (unless or until there somehow ceases to be any record of the outcome of that measurement in the position of even so much as a single subatomic particle anywhere in the universe; and *that* can presumably almost *never* come to pass, if the outcome of the

measurement ever gets recorded in anything macroscopic) the measured system (or rather: the positions of all of the particles which
make up that system) will subsequently evolve just as if that
system's wave function has been *collapsed,* by the measurement,
onto an *eigenfunction* (the one corresponding to the measured
eigenvalue) of the measured observable, even though as a matter of
fact it *hasn't* been; and it also entails that the *probabilities* of those
"collapses" will be precisely the familiar quantum-mechanical
ones.

Mentality

But there are interesting questions (which I want to just set up here,
and then leave to the reader) about whether or not all that turns
out to be enough.

Consider (for example) whether Bohm's theory guarantees that
every sort of measurement whatsoever even *has* an outcome.
Bohm's theory does a good deal *better* at that than any theory of
the collapse of the wave function can (that's what we found out in
note 5); but there are problem cases for Bohm's theory too.

Here's a science-fiction story (along the lines of the story about
John, in Chapter 5) about one of those.

This story is going to involve a device (like the one depicted in
figure 7.14) for producing a correlation between the *hardness* of a
certain microscopic particle P and the hardness of an incoming
electron (which is slightly different from what happens in the story
in Chapter 5); a sort of measuring device for the hardness of
electrons in which the "pointer" is the *hardness* of P. Here's how
the device works: if the device starts out in its ready state, and if
(say) a hard electron is fed through the device, then the hardness
of that electron is unaffected by its passage through the device, and
the device is unaffected too, except that P ends up, once the electron
has passed all the way through, *soft;* and things work similarly (or
rather, analogously) if a *soft* electron is fed through the device, if
the device is initially in its ready state (in that case P ends up *hard*).
Once the hardness of the electron gets correlated to the hardness
of P, then P's hardness can be measured, if we want to, by means

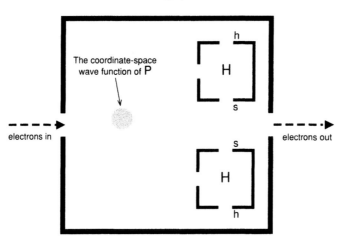

Figure 7.14

of either one of the two little hardness boxes (note that one of them is right-side-up and the other is up-side-down, but of that more later) on the right-hand side of the device.

A device like the one I just described is (according to the story) sitting in the middle of John-2's brain, and the particular way in which that device is now hooked up to the rest of John-2's nervous system makes him function as if his *occurrent beliefs* about the hardness of electrons that happen to pass through it are determined *directly* by the hardness of P (just as the occurrent beliefs of the *original* John about the hardness of those sorts of electrons were determined directly by the *position* of *his* P).

Here's what that means: Suppose that John-2 is presented with an electron which happens to be hard and is requested to ascertain what the value of the hardness of that electron is. What John-2 does (just as the original John did) is to take the electron into his head through his right door, pass it through his surgically implanted device (with the device initially in its ready state), and then expel it from his head through his left door. And when that's all done (that is, when P is soft but the value of the hardness of the electron is not yet recorded anywhere in John-2's brain *other* than in the hardness of *P*), John-2 announces that he is, at present, consciously aware of what the value of the hardness of the electron is, and that he would be delighted to tell *us* what that value is, if we would like to know.[14]

Now, the way that John-2 behaves, subsequent to all that, in the event that we *do* ask him to tell us what that value is, is (according to this story) as follows:

The rules of the game (to begin with, to keep things simple) are that John-2 can tell us about that either *verbally* or *in writing* (or in both ways, but of that more in a minute).[15]

And what the story stipulates is that in the event that we ask John-2 to tell us *verbally* what the value of the hardness of the electron is, then what he does (what he's programmed to do, what he's wired up to do) is to pass *P* through the upper hardness box (the one that's right-side-up), which is equipped with *P* detectors at its two exit apertures, which are in turn connected with certain of John-2's neurons, which are themselves ultimately connected with his mouth muscles and his throat muscles and his tongue muscles in such a way as to end up generating (in the case where John-2 is initially presented with a *hard* electron) the utterance "hard" (and of course in the event that John-2 is initially presented with a *soft* electron, then these same procedures and these same connections will end up producing the utterance "soft").

And (on the other hand) the story stipulates that in the event that what we ask John-2 to do is to report *in writing* what the value of the hardness of the electron is, *then* what he does (what, once again, he's wired up to do) is to pass *P* through the *lower* hardness box (the one that's *upside-down*) which is similarly (or rather, analogously) equipped with *P* detectors and hooked up to the rest of

14. There was a good deal of talk in Chapter 5 (which the reader would do well to bear in mind here too) about precisely how we ought to *take* announcements like that.

15. Of course, there are almost certainly going to be any number of other ways (*besides* speaking or writing) in which anybody like John-2 could manage to communicate information like that, but those other ways need not concern us right now; all that will turn out to be important for our present purposes is that there can be at least two ways for him to communicate it.

John-2's brain in such a way as to end up generating (in the case where John-2 is initially presented with a *hard* electron) the *inscription* "hard" (and of course, as above, in the event that John-2 is initially presented with a *soft* electron, then these same procedures and these same connections will end up producing the inscription "soft").

Good; now (and here comes the interesting part of the story) consider how John-2 is going to behave in the event that he is presented with (say) a *white* electron and requested to measure the hardness of that electron, and to remember the outcome of that measurement. When John-2 is done with all that (that is: when the electron has passed all the way through John-2's head, as depicted in figure 7.15; and when John-2 informs us that he now knows what the value of the hardness of that electron is, that he is now consciously *aware* of what the value of the hardness of that electron is, and that he can *tell us* that value, if we like, either verbally or

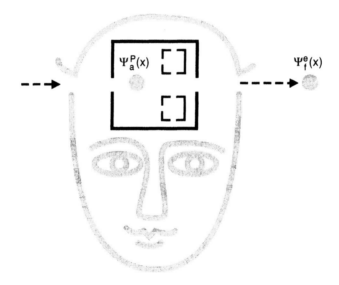

John-2

Figure 7.15

in writing), then (as the reader can easily confirm, since we've already had a good deal of practice with this sort of thing) the state of the composite system that consists of P and the measured electron is going to be:

$$(7.14) \qquad \tfrac{1}{\sqrt{2}}(|\text{soft}\rangle_P|\psi_a(x)\rangle_P|\text{hard}\rangle_e|\psi_f(x)\rangle_e \; - \; |\text{hard}\rangle_P|\psi_a(x)\rangle_P|\text{soft}\rangle_e|\psi_f(x)\rangle_e)$$

where $\psi_a(x)$ is a wave function which is nonzero only in the region a in figure 7.15, and $\psi_f(x)$ is a wave function which is nonzero only in the region f in figure 7.15. And of course the state in (7.14) is precisely the same state as the one in (7.10), which we discussed in considerable detail above.

And so (and all of what follows now can just be read off from what we found out before about (7.10)) if we were to ask John-2 to tell us, verbally (when (7.14) obtains), what the hardness of the measured electron is, then he would utter either "hard" or "soft,"[16] and of course the procedures that go on inside of John-2's head which *lead up* to one or the other of those two utterances will produce an *effective collapse* of the wave function in (7.14) onto (respectively) either its first or its second term, and so whichever of those two utterances John-2 ends up making will necessarily be *confirmed* to be correct by *any* subsequent measurement of the hardness of the electron *itself*.

And if, instead, we were to ask John-2 to tell us *in writing* (when (7.14) obtains) what the hardness of the measured electron is, then he would *write down* either "hard" or "soft."[17] The procedures that go on in John-2's head which lead up to one or the other of those *inscriptions* will produce the same sort of effective collapse as above, and so whichever of those two inscriptions John-2 ends

16. Note, by the way, that there is invariably going to be an objective and unambiguous matter of fact about *which* of those two utterances John-2 *makes*, since the act of uttering "soft" can be distinguished from the act of uttering "hard" in terms of the positions of all sorts of things (certain parts of John-2's tongue, for example, or his lips, or his throat).

17. And here again, for the same sorts of reasons, there will invariably be an objective matter of fact about which one of those he *does* write down.

up making will *also* necessarily be confirmed to be correct by any subsequent measurement of the hardness of the electron itself.

And moreover, if we were to ask John-2 for a verbal report on the hardness of the measured electron, when (7.14) obtains, and if, once that report is produced, we were to ask him for a *written* report on the hardness of that same electron, then the content of that written report will invariably coincide with the content of the earlier verbal report; and the content of any *second* verbal report on the hardness of that electron will also invariably coincide with the content of that first one; and the content of any later *written* report on that hardness will *also* invariably coincide with the content of any earlier one, and so on.

And so whatever plausible observable or functional criteria there may be for having a belief, or for having a true belief, or for having a justified true belief about the hardness of that measured electron, *all* of those criteria will plainly be satisfiable by anybody wired up as John-2 is, when a state like the one in (7.14) obtains.

And yet (and this is the punch line) if Bohm's theory is right, then John-2 can't *possibly* be a genuine "knower" of the hardness of the measured electron, when (7.14) obtains.

Here's why: Suppose (for example) that P happens to be in the upper half of region *a*, when (7.14) obtains, and suppose that John-2 is first requested to produce a verbal report about the hardness of the electron. Then the content of that report, and of any subsequent report, of either kind, about that hardness is going to be that the electron is *soft*. But in the event that John-2 is in precisely the condition described above and he is first requested to produce a *written* report about the hardness of the electron, *then* the content of *that* report, and of any subsequent report, of either kind, about that hardness is going to be that the electron is *hard*. And of course all of that will get *reversed* in the event that P happens to be in the *lower* half of region *a*, when (7.14) obtains.

And so, when (7.14) obtains (that is: before any report has been requested of John-2, but when John-2 claims to be a *knower* of the hardness of the electron, and when John-2 is in fact in a position to produce a *report* about that hardness, of *either* sort, which will with certainty be confirmed by any subsequent measurement of that hardness), John-2's *dispositions* are *completely unlike* those of any genuine rational *knower* of that hardness; since the *content* of whatever reports John-2 produces about that hardness will depend (as a matter of *fact,* but *not* in any *observable* way) on what *sort* of report is requested *first.*

The Incommensurability of Bohm's Theory and Many–Minds Theories

The business of comparing the empirical contents of Bohm's theory to the empirical contents of a *many-minds* theory turns out to be very peculiar.

To begin with, if *either one* of those two theories happens to be *the true theory of the world,* then the question of *which one* of them is the true one will be undeterminable by any imaginable sort of empirical evidence. It's pretty clear how that works: Each one of those two theories entails (as I've already mentioned) that the linear dynamical equations of motion are always exactly right, and each one of them entails that the probabilities of the outcomes of all of the sorts of experiments that *have* any outcomes (we'll talk more about that soon) are precisely the ones laid down in principle D of Chapter 2, and so (as a matter of fundamental principle) *there can't be any purely experimental means of deciding between them.*[18]

---

18. Any theory of the *collapse of the wave function,* by the way, is clearly going to *differ* from Bohm's theory and from a many-minds theory (and presumably from *any* theory in which there *isn't* any such thing as a collapse of the wave-function), in terms of its predictions about the outcomes of certain experiments.

Here's how that will work. Any theory of the collapse of the wave function (just in virtue of *what it is* to be a theory of the collapse of the wave function) is going to entail that there are certain superpositions of macroscopically different states of the world which, whenever they arise, collapse (more or less immediately) onto *one or another* of those states. Take any such theory (let's call it *T*). Design a hardness measuring device for which the "indicates-that-hard" state and the "indicates-that-soft" state are (as it were) macroscopically different *insofar as T is concerned* (that is: design the device in such a way as to guarantee that *T* entails that a state like the one in equation (5.2) will more or less immediately collapse onto the state in equation (5.1)). Prepare the device (either genuinely or effectively) in its ready state.

So, suppose it were to turn out that what our empirical experience entails is that the dynamical equations of motion are always exactly right (which is how I figure it probably will turn out, once the results are in); and suppose it were also to turn out that there aren't any purely theoretical reasons why one or the other of these two theories is somehow manifestly out of the running (which is maybe a little less certain). What that would mean is that questions about the structure of space and time, and questions about whether or not the world is deterministic (which were supposed to be the two central questions of the physics of this century, and which both happen to be questions on which these two theories radically disagree with one another), are the kinds of questions which there can't ever be scientific answers to. Period.

But there's something else that's weird too.

Consider the question (the one that I postponed a few paragraphs back) of *which* experiments *do* have any outcomes. Bohm's theory and many-minds theories disagree about that.

Consider, for example, a measurement of the hardness of a black electron that gets carried out by an automatic device with a pointer, like the one in figure 5.1. On Bohm's theory, there will be a determinate matter of fact about where that pointer ends up, and where that pointer ends up will be the sort of thing that counts as a piece of absolutely raw empirical data, the sort of thing (that is) that physics is fundamentally in the business of making predictions about. And of course on a *many-minds* theory, there *won't* be any determinate matter of fact about where that pointer ends up, and (consequently) where that pointer ends up can't *possibly* be the sort of thing that counts as a piece of absolutely raw empirical data, and (consequently) where that pointer ends up can't *possibly* be the

---

Feed a black electron into it. Let the interaction between the device and the electron run its course. Then measure the observable of the composite system consisting of the device and the electron that we talked about on pages 87–92, the one we called zip − color there, the one that's extremely hard (but not impossible) to measure. If T is true, then there will be nonzero probabilities of a number of different outcomes of the zip − color measurement; but if Bohm's theory is true, or if a many-minds theory is true, *then* the outcome of that measurement will invariably and with certainty be *zero*. And that's that.

sort of thing that physics is in the business of making any *predictions* about. And of course the conviction of any adherent of *Bohm's* theory to the effect that there *is* some matter of fact about where that pointer ends up will be perfectly explicable in the context of a *many-minds* theory as a simple *delusion*. And (consequently) there *can't* be any means of *resolving* that dispute.

And consider a measurement of the hardness of a black electron that gets carried out by a sentient observer whose brain is wired up as John-2's brain is. On a many-minds theory, there will be a determinate matter of fact about what each one of John-2's minds ends up thinking about the hardness of that electron, and those *thoughts* will be the sorts of things that count as pieces of absolutely raw empirical data and as the sorts of things that physics is fundamentally in the business of making predictions about. And of course on *Bohm's* theory, there *won't* be any matter of fact about what John-2 ends up thinking about the hardness of that electron, and (consequently) what he ends up thinking about that can't *possibly* be the sort of thing that counts as a piece of absolutely raw empirical data, or as the sort of thing that physics is fundamentally in the business of making predictions about. And of course the conviction of any adherent of a *many-minds* theory to the effect that there is some matter of fact about what any one of John-2's minds ends up thinking about that will likewise be perfectly explicable (because of the stuff we talked about in the first section of Chapter 6), in the context of *Bohm's* theory, as a delusion. And (consequently) there can't be any means of resolving *that* dispute either.

And so the upshot of all this is that it doesn't capture what's going on to say of these two theories that (in virtue of the impossibility of experimentally *telling them apart*) they're empirically *equivalent* to one another. What these two theories *are* is (in an extraordinarily radical way) empirically *incommensurable* with one another: What a many-minds theory takes physics to be ultimately about is *what observers think;* and it entails that there will frequently not even be *matters of fact* about *where things go.* And what *Bohm's* theory takes physics to be ultimately about (as I mentioned before) is *where things go;* and *it* entails that there will sometimes not even be matters of fact about *what observers think.*

Of course, there may happen to be sentient observers in the world whose thoughts supervene entirely on the *positions* of things. *Human* observers (the ones whose brains are wired up in the *natural* way, that is) are probably (most of the time) like that. And of course Bohm's theory and many-minds theories will *agree* that there are determinate matters of fact about the outcomes of any measurements that get carried out by observers like *that;* and (moreover) they will agree (as I mentioned above) that the *probabilities* of those outcomes will be precisely the ones laid down in principle D of Chapter 2. But that kind of thing (if it happens) will amount to a rather special case.

··· 8 ···

Self–Measurement

If there isn't any such thing as a collapse of the wave function (which is what smells right to me, and which the reader ought to be persuaded, by now, is at least worth considering), then there are stories about what's physically possible that have interesting surprises in them.

And I want (by way of finishing this book up) to tell one.

Let me tell it first in the bare-theory language.

Suppose that a competent observer named h carries out a measurement of the color of an electron that's initially hard. When that's done (as we've seen a number of times already) the physical state of h and of her measuring apparatus and of the electron is going to be

$(8.1) \qquad 1/\sqrt{2}(|\text{believes } e \text{ black}\rangle_h|\text{"black"}\rangle_m|\text{black}\rangle_e$

$\qquad\qquad + |\text{believes } e \text{ white}\rangle_h|\text{"white"}\rangle_m|\text{white}\rangle_e)$

Let's call the state in (8.1) $|A\rangle_{h+m+e}$. And bear in mind that $|A\rangle_{h+m+e}$ is necessarily going to be an eigenstate of some complete set of physical observables, since every quantum state necessarily is. Let's call those observables (in the particular case of $|A\rangle_{h+m+e}$) $\{Q\}$, and let's call the associated eigenvalues $\{q\}$.

Now, suppose that there happens to be a *second* observer around (call him $h2$); and suppose that at the conclusion of h's measurement of the color of e (when the state of h and m and e is $|A\rangle$), $h2$ *measures* $\{Q\}$. He carries out that measurement (in the usual way) by means of a $\{Q\}$ measuring device (which is a device which can

180

necessarily be constructed, if quantum mechanics is right)[1] which interacts with h and m and e and which *records* the values of $\{Q\}$ in (say) the positions of some set of pointers, and which $h2$ subsequently *looks at* in order to ascertain what the values of $\{Q\}$ *are*.

When that's all done (since the result of this measurement is with certainty going to be $\{Q\} = \{q\}$), things are going to look like this:

(8.2) $|\text{believes } \{Q\} = \{q\}\rangle_{h2}|\text{indicates } \{Q\} = \{q\}\rangle_{mQ}|A\rangle_{h + m + e}$

What's *going on* in this state is that $h2$ is (as it were) in possession of something like a *photograph* (in his $\{Q\}$ measuring device) of the full superposition of h's brain states which constitutes $|A\rangle_{h + m + e}$.

Of course, there isn't anything particularly surprising in *that*. In so far as $h2$ is concerned, after all, h, whatever else she is, is a physical system out there in the external world; and so h ought in principle to be no less susceptible of being in superpositions of these sorts of brain states, and no less susceptible of being *measured* to be in superpositions of these sorts of brain states, than (say) an electron is susceptible of being in, and of being measured to be in, superpositions of *hardness* states. But we're not at the end of the story yet.

Suppose that now (when the state in (8.2) obtains) $h2$ tells h how the $\{Q\}$ measurement came out. (Note, by the way, that $h2$'s function in this story is merely to be an *indicator*, to h, of the outcome of the $\{Q\}$ measurement; and so the story we're in the middle of here can just as well be thought of as the story of a measurement of the $\{Q\}$ which h carries out, with the help of the appropriate *instruments*, on *herself*). Well, the way things are going to end up when *that's* done (given the linearity of the equations of motion) is like this:

---

1. Of course, the construction of a device like that is in general going to be extraordinarily difficult, for the reasons described on pages 88–92. But that need not concern us right now. We're just telling a science-fiction story here; all that's important is that it be physically *possible*.

(8.3) $|\text{believes } \{Q\} = \{q\}\rangle_{h2}|\text{indicates } \{Q\} = \{q\}\rangle_{mQ}$
$\times \ 1/\sqrt{2}\{|\text{believes } e \text{ black and } \{Q\} = \{q\}\rangle_h|\text{"black"}\rangle_m|\text{black}\rangle_e$
$+ |\text{believes } e \text{ white and } \{Q\} = \{q\}\rangle_h|\text{"white"}\rangle_m|\text{white}\rangle_e\}$

And *this* is going to turn out to be a really curious state of affairs.[2]

＊   ＊   ＊

2. Let's say a bit more about the structure of h's brain.

Suppose that h stores her belief about the hardness of the electron in a certain set of memory elements called a; and that she stores her beliefs about the values of the $\{Q\}$ in a certain set of memory elements called b. And let's refer to all of the rest of h as h^*. And let's assume (just to keep things simple) that the state of h^* throughout the story we've been telling here is some particular state called $|\&\rangle_{h^*}$. Then we can write:

$|\text{believes } e \text{ black and } \{Q\} = \{q\}\rangle_h = |\&\rangle_{h^*}|\text{"black"}\rangle_a|\text{"}\{Q\} = \{q\}\text{"}\rangle_b$

and

$|\text{believes } e \text{ white and } \{Q\} = \{q\}\rangle_h = |\&\rangle_{h^*}|\text{"white"}\rangle_a|\text{"}\{Q\} = \{q\}\text{"}\rangle_b$

And if you write out the whole state in (8.3) in terms of states of a and b and h^*, it looks like this:

$|\text{believes } \{Q\} = \{q\}\rangle_{h2}|\text{indicates } \{Q\} = \{q\}\rangle_{mQ}$
$\times \ 1/\sqrt{2}\{|\&\rangle_{h^*}|\text{"black"}\rangle_a|\text{"}\{Q\} = \{q\}\text{"}\rangle_b|\text{"black"}\rangle_m|\text{black}\rangle_e$
$+ |\&\rangle_{h^*}|\text{"white"}\rangle_a|\text{"}\{Q\} = \{q\}\text{"}\rangle_b|\text{"white"}\rangle_m|\text{white}\rangle_e\}$

or (equivalently) like this:

$|\text{believes } \{Q\} = \{q\}\rangle_{h2}|\text{indicates } \{Q\} = \{q\}\rangle_{mQ}|\&\rangle_{h^*}|\text{"}\{Q\} = \{q\}\text{"}\rangle_b$
$\times \ 1/\sqrt{2}\{|\text{"black"}\rangle_a|\text{"black"}\rangle_m|\text{black}\rangle_e + |\text{"white"}\rangle_a|\text{"white"}\rangle_m|\text{white}\rangle_e\}$

And now we can clear something up. What I said about the $\{Q\}$ in the text was that they form a *complete set* of observables of the composite system that consists of h and m and e. But that can't be exactly right, because any specification of the values of any complete set of observables of any given system is supposed to *uniquely* pick out some *particular state* of that system (that's what "complete" is supposed to *mean*), and yet the state of h and m and e in (8.1) obviously isn't the same as the state of h and m and e in (8.3) (since in (8.3) h has *heard* about the values of the $\{Q\}$, and in (8.1) she hasn't), and nonetheless (8.1) and (8.3) are both supposed to be eigenstates of the $\{Q\}$, with (in *both* cases) the eigenvalues $\{q\}$.

Here's what's exactly right: What the $\{Q\}$ form a complete set of observables of

Suppose that (8.3) obtains, and suppose that some *third* observer (call her $h3$) instructs $h2$ to go out and measure the values of the $\{Q\}$ and then to ask h what *she* takes the values of the $\{Q\}$ to be, and see if h is right about them and report back. When that's done, $h2$ will report (with certainty, if he's a competent observer) that h is right.

And suppose that (8.3) obtains and that $h3$ instructs $h2$ to go out and measure the color of the electron and then to ask h what she takes the color of the electron to be, and see if she's right about it, and report back. When that's done, $h2$ will report (with certainty, if he's a competent observer) that h is right.

And suppose that (8.3) obtains and that $h3$ instructs $h2$ to measure the values of the $\{Q\}$ *and* to measure the color of the electron, and then to ask h what she thinks about *both* of those things and report back. When *that's* done, $h2$ will report (with certainty, if he's a competent observer) that h is right; $h2$ will report (that is) that h has *correct beliefs* (when (8.3) obtains) about *both* $\{Q\}$ *and* the color of the electron.

So what's going on in (8.3) is that h knows (effectively) the value of the color of the electron, and h *also* knows *(genuinely)* the values of $\{Q\}$, even though $\{Q\}$ and the color of the electron are quantum-mechanically *incompatible* with one another.[3]

And of course that amounts to a direct violation of the familiar quantum-mechanical uncertainty relations.[4]

٭ ٭ ٭

(that is: what the state $|A\rangle$ refers to) isn't h and m and e, it's a and m and e; and the state of a and m and e in (8.1) *doesn't* differ from the state of a and m and e in (8.3); and (8.1) and (8.3) *are* both eigenstates of the $\{Q\}$ with the eigenvalues $\{q\}$.

3. That is: the state $|\{Q\} = \{q\}\rangle$ is a *superposition* of states associated with *different* eigenvalues of the color of the electron.

4. And so something has got to be *wrong* with the familiar quantum-mechanical uncertainty relations (and with a good deal else of what's been said about measurements in this book, too) on theories in which there isn't any such thing as a collapse of the wave function. Let's see what that is.

Suppose that h carries out a measurement of some observable called O of some physical system called S. And suppose that when that measurement is done h stores

Let's look into that some more.

Suppose that (8.3) obtains and that (consequently) h simultaneously knows the color of the electron (effectively) and the values of $\{Q\}$ (genuinely); and suppose that $h2$ decides that *he'd* like to know those two things simultaneously too. Well, he's already half-

---

her memory of the value of O in some particular observable of *her brain* called $M(O)$.

If we want to describe quantitatively how well h has *succeeded* in ascertaining what the value of O is, in the course of a measurement like that, we can define an observable called $E(O)$ of the composite system $S + h$ which measures the *error* in h's belief about O, viz.:

$$E(O) = M(O) - O$$

And what it will *mean* (in this quantitative language) to say of h that she either genuinely or effectively *knows the value of O is just that the state of $S + h$ is an eigenstate of $E(O)$* with eigenvalue 0. And so (on any theory in which there isn't any such thing as a collapse of the wave function) the question of whether or not the values of any *two* particular observables A and B can *simultaneously* be genuinely or effectively known to h will reduce to the question of whether or not there can be any state of the world which is an eigenstate of $E(A)$ with eigenvalue 0 and which is *also* an eigenstate of $E(B)$ with eigenvalue 0, and of course *that* question will reduce (in turn) to the question of whether or not $E(A)$ and $E(B)$ are quantum-mechanically *compatible* with one another, and what we've discovered in this chapter (and what's been *overlooked* in all of the *preceding* chapters of this book, and in all of the *standard* discussions of the theory of measurement) is just that *that* question *isn't* always precisely the same as the question of whether or not A and B *themselves* are quantum-mechanically compatible with one another.

But we're getting a little ahead of ourselves. Let's take it one step at a time.

Suppose (to begin with) that A and B are both observables of physical systems which are *external* to h's *brain*. Then A and B will each necessarily be quantum-mechanically compatible with both $M(A)$ and $M(B)$ (since A and B will each be observables of physical systems *other than* the one that both $M(A)$ and $M(B)$ are observables of). Moreover, if h is to be able to *report* her belief about the value of A without (in the process) disrupting her memory about the value of B, and if she is to be able to report her belief about the value of B without (in the process) disrupting her memory about the value of A (and the possibility of doing all *that* is presumably just a part of what it *means* to have beliefs about the value of A and the value of B), then $M(A)$ and $M(B)$ will also have to be quantum-mechanically compatible with *one another*.

way there, since (if (8.3) obtains) he already knows the values of {Q}. All he needs to do now is to measure the color of the electron, or (just as good) to ask *h* what the color of the electron is, or whatever. But the trouble is that doing any of those things is necessarily going to produce a state that's *nonseparable* between the electron and the color measuring device and *h and h*2, a state in which some observable of the brain of *h*2 is *correlated* with the value of the color of the electron; and so *that* state (the one that things *end up* in) can't *possibly* be an eigenstate of the {Q} (since the {Q} are operators of the composite system that consists just of the electron and the color measuring device and *h*, and *not h*2); and so (once that's done) *h*2's beliefs about the values of {Q} will necessarily be *false*.

And of course that's just a special case of something more general, which is that if any physical observable whatever gets correlated with the color of the electron (except, of course, for the ones that are *already* correlated with it when (8.3) obtains), then (once that correlation is established) the state of the world can't possibly, any longer, be an eigenstate of the {Q}. And so no observer in the world other than *h herself* can *ever* be in a position (no matter *what*

And it can be shown to follow from all *that* (see Albert, 1983) that $E(A)$ and $E(B)$ can be compatible with one another only in the event that A and B are compatible with one another.

And so the dynamical equations of motion by themselves will entail that in any of the circumstances that were imagined in any of the *previous* chapters of this book, and in any of the circumstances that get imagined in any of the *standard* treatments of the theory of measurement (in all of which observers carry out observations only on things *external to themselves*), the familiar quantum-mechanical uncertainty relations will invariably apply in precisely the familiar way. And of course it's *just that* that makes it *entertainable* that those equations are (as a matter of fact) *always exactly right*.

But what's going on in the state in *(8.3)* is an altogether different matter. The state in (8.3) is an eigenstate of E(hardness of the electron) with eigenvalue 0, and it is *also* an eigenstate of $E(\{Q\})$ with eigenvalue 0, and nonetheless the hardness of that electron is *quantum-mechanically incompatible* with the {Q}, and the *possibility* of all that hinges on the fact that M(hardness of the electron) is *also* quantum-mechanically incompatible with the {Q}.

the quantum state of the world is) to simultaneously know (either genuinely or effectively) the values of the color of the electron and of the $\{Q\}$.[5]

＊　＊　＊

Let's talk about (8.3) over again in the *many-minds* language.

One of the things that's going on when (8.3) obtains (in *this* language) is that half of h's minds believe that the electron is black and half of them believe that the electron is white.

And (moreover), when (8.3) obtains, each one of h's minds is effectively right about the color of the electron. And the sense of *effective* here, remember, is the *stronger* one, the one appropriate to *many-minds* talk: each one of h's minds (when (8.3) obtains) knows, genuinely, with certainty, what it would perceive if *another* measurement of the color of the electron were now to be carried out.

And (moreover), when (8.3) obtains, all of h's minds *also* know (genuinely) that $\{Q\} = \{q\}$. And so the minds that effectively know that the electron is black are *simultaneously* aware (in virtue of the outcome of a *measurement*) of the existence of a branch of the wave function of their own brain associated with the belief that the electron is *white,* and the minds that effectively know that the electron is *white* are simultaneously aware (in virtue of the outcome of the *same* measurement) of the existence of a branch of the wave

5. What's special about h, in the story we've been telling here, is (of course) that *she's* the one to whom the $\{Q\}$ refer.

But we can obviously tell a very similar story about h2. *That* story will involve some observables (call them $\{Q'\}$) which refer to $h2$ in just the same way as the $\{Q\}$ refer to h; and what will *emerge* from that story is that h2 *can* potentially simultaneously know the values of the color of the electron and the $\{Q'\}$ and that h can't, and that everybody else can't either.

And so there are going to combinations of things that any particular observer can (in principle) simultaneously know about the world, and that nobody else can simultaneously know about it; what those things are will depend on the identity of the observer in question, and some of those things will invariably be things that are about that observer herself.

function of their own brain associated with the belief that the electron is *black*.[6]

In many-*worlds* talk (if that kind of talk made sense), this sort of thing would amount to an awareness (in virtue of the outcome of a measurement) of the existence of *other worlds*.

Let's talk about (8.3) over again in *Bohm's* language; and let's suppose (just to keep us out of the sort of trouble we ran into in Chapter 7) that *h* stores her memories about the color of the electron and the values of the {Q} in the *positions* of some particles in her brain.

Bohm's theory is completely deterministic. There are already objective physical matters of fact, when (8.3) obtains (on Bohm's theory), about how any subsequent measurements of the color of the electron or of the {Q} or of both of them are going to come out. Moreover, when (8.3) obtains, *h knows, genuinely, with certainty, what those facts are*.[7] And no *other* observer in the world

6. Consider (by the way) how *h*'s minds *evolve*, on this picture, in the course of the measurement of the {Q} that leads from (8.1) to (8.2). The principle we've been using so far (remember) is that each individual one of *h*'s minds always evolves just as if that mind's present beliefs about the quantum state of the world were actually true. And what *that* will entail (even though the {Q}-measurement produces no change whatever in the physical state of *h*) is that half of those of *h*'s minds that believed the electron "white" before the {Q}-measurement will believe it is "black" *after* it, and that half of those of *h*'s minds that believed the electron "black" before the {Q}-measurement believe it is "white" after it. And so (if this is really how things work) once the {Q}-measurement is over, *h*'s *memories* of what she *thought* about the color of the electron *before* that measurement won't be reliable ones.

Of course, none of this will undercut what's just been said in the text: when (8.3) obtains, the beliefs of every one of *h*'s minds about the outcome of any *upcoming* measurement of the color of the electron (no matter *how* those beliefs actually *came into being*) will invariably be effectively *true*. And (needless to say) the beliefs of every one of *h*'s minds about any upcoming measurement of the {Q}, when (8.3) obtains, are going to be *genuinely* true.

7. That is: when (8.3) obtains, either *h* believes that any subsequent measurement of the color of the electron will have the outcome "black" or she believes that any subsequent measurement of the electron will have the outcome "white"; and whichever one of those she *does* believe is (as a matter of *objective physical fact*) *true*. And of course when (8.3) obtains *h* also genuinely knows that {Q} = {q}.

can *ever* be in a position to know, all at the same time, what those facts are.

Think about how curious that is. Consider (for example) the following game: Suppose that (8.3) obtains, and suppose that *h* consents to allow some future act of hers to be determined (in accordance with some fixed decision table) by the results of some *upcoming* measurements of the {*Q*} and of the color of the electron.[8] On Bohm's theory, there is, right now (that is: *before* those upcoming measurements get carried out) an objective physical matter of fact about what that future act of *h*'s is going to be; and (moreover) *h* now *knows, with certainty,* what that act is going to be; and (moreover) no *other* observer in the world (no matter how adept they may be at measuring or calculating) can *possibly* know (right now) what that act is going to be.

And so *h*, under these sorts of circumstances (even though the complete physical theory of the world here is a *deterministic* one), has what you might call an inviolably private will.[9]

* * *

8. Here's what I mean: Suppose that (8.3) obtains now, and the rules of the game are that (say) one hour from now the {*Q*} will be remeasured and then (right after that) the color of the electron will be remeasured; and then (once those remeasurements are done) *h*'s next act will be determined (in accordance with some fixed decision table) by how those remeasurements come out.

9. Consider, on *this* theory, how *h*'s mental state evolves in the course of the {*Q*}-measurement that leads from (8.1) to (8.2). This turns out to be a very different business here than it was on the many-minds picture.

What's crucial is that the {*Q*}-measurement produces *no changes whatsoever* in the overall wave function of *h* and *m* and *e*, and (consequently) it produces no *probability currents* in the wave function of *h* and *m* and *e*, and (consequently) it produces no changes in the *positions of the particles* that *constitute h* and *m* and *e*, and (consequently) it will produce no evolution of *h*'s mental state *at all.** And so, on *this* theory, *h*'s memory (when (8.2) obtains) of what she thought about the color of the electron before the {*Q*}-measurement will be *perfectly* reliable; and *h*'s beliefs (when (8.3) obtains) about the outcomes of any upcoming measurements of the color or of the {*Q*} or of *both* will all be perfectly right too.

*Given these differences, the relationship between Bohm's theory and many-minds theories is even more complicated than we understood it to be at the end of

And so the mental lives of quantum-mechanical observers who could arrange to carry out these sorts of measurements on their own brains would perhaps be unimaginably (for us) rich.

Chapter 7. It turns out (in particular) that there can be explicit disagreements between these two theories about the probabilities that certain minds will have certain thoughts even under circumstances in which *both* theories *agree* that there will with certainty be matters of fact about *what* those thoughts *are!* Nonetheless (as the reader will see if she thinks the business through, or if she refers to the recent paper on these matters by Hilary Putnam and myself (1992), even *these* sorts of disagreements will prove impossible to settle by means of any kind of an *experiment*.

Appendix

The Kochen–Healy–Dieks
Interpretations

One more small tradition of attempts to solve the measurement problem ought to get mentioned here. They don't work,[1] but (since that isn't generally known yet) they've gotten a good deal of attention lately.

These attempts (which are due to Kochen, 1985, Dieks, 1991, and Healy, 1989, and which are all more or less the same in their essentials) I'll refer to here as *KHD* interpretations of quantum mechanics. They all respond to the measurement problem, just as *Bohm's* theory does, by insisting that the dynamical equations of motion are always exactly right, and by *denying* that an observable pertaining to a physical system has a determinate value only in the event that the quantum state of that system happens to be an *eigenstate* of that observable.

What *Bohm's* theory says, of course, is that there are certain particular observables pertaining to physical systems (namely: the positions of particles) which *invariably* have determinate values, no matter *what* the overall wave function of the world is.

The KHD interpretations are a little more sophisticated than that. On the KHD interpretations, the identities of those observables which have determinate values (over and above the ones whose values are determined, in accordance with the standard way of thinking, by the overall wave function of the world) can *vary* from moment to moment; and those identities *depend* on what the

---

1. The arguments which *show* that they don't work, which is what this appendix will mainly be about, were first presented in a couple of recent papers by Barry Loewer and myself (see, for example, Albert and Loewer, 1991).

overall wave function of the world is, and the particular *way* in which they depend on what that overall wave function is (that is: the explicit *rules* whereby they depend on what that overall wave function is) are cooked up so as to guarantee (apparently) that measurements always have outcomes.

Let's set things up.

Suppose that the dynamical equations of motion are always exactly right, and consider a measurement of (say) the hardness of an electron whose initial state is

(A.1) $a|\text{hard}\rangle_e + b|\text{soft}\rangle_e$

When the measurement interaction is over, the state of the composite system consisting of the measuring device and the measured electron is

(A.2) $a|\text{``hard''}\rangle_m|\text{hard}\rangle_e + b|\text{``soft''}\rangle_m|\text{soft}\rangle_e$

Now, the state in (A.2) can of course be written down in any number of different *bases;* but it happens that the basis that it *has* been written down in, in (A.2), has a very special feature. If $a \neq b$, then the basis that (A.2) happens to be written down in is (according to a theorem of Schmidt) the *only* basis in which that state will take the "bi-orthogonal" form:

(A.3)

$$\sum_i c_i|a_i\rangle_m|b_i\rangle_e$$

with $\langle a_i|a_j\rangle = \langle b_i|b_j\rangle = 0$ unless $i = j$

O.K. Here's how the KDH interpretations work. The rules (over and above principles A, B, and C of Chapter 2) are as follows:

Rule 1. If some physical system S can be divided up (in any way at all) into two subsystems S_1 and S_2, and if the overall quantum state of S is $|Q\rangle$, and if the unique bi-orthogonal representation of $|Q\rangle$ in terms of states S_1 and S_2 is $\Sigma_i c_i|a_i\rangle_1|b_i\rangle_2$, where the $|a_i\rangle$ are eigenstates of an observable called A and the $|b_i\rangle$ are eigenstates of an observable called B, then S_1 has a determinate value of observable A and S_2 has a determinate value of observable B.

Rule 2. If S is in the state $|Q\rangle$, then the probability that $A = a_i$ on S_1 and $B = b_i$ on S_2 is c_i^2.

And that (at first sight) seems to work extraordinarily well, because rule 1 entails that when states like (A.2) obtain, there will invariably be some determinate matter of fact about what the hardness of the electron is and about what m indicates the hardness of the electron to be. And as a matter of fact, it's easy to see that rule 1 will entail that at the conclusion of any interaction whatever between a good measuring device (which starts out in its ready state) and a measured system (which can start out in any state at all) there will be some determinate matter of fact about the value of whatever observable of the measured system it is that that device is designed to measure, and also about what the measuring device *indicates* about that value. And of course rule 2 will guarantee that the *probability* of that one such value as opposed to another obtaining, or being indicated, will be the *right* one. And all that gets done (just as in Bohm's theory) without a collapse of the wave function.

But there are difficulties. Let's run through (in order of increasing seriousness) three of them.

1. In the event that any two of the c_i's in (A.3) are *equal* to one another, there will be an *infinity* of *different* bi-orthogonal decompositions of the state in question; and so in cases like that rule 1 will entail that an *infinity* of mutually incompatible observables of S_1 and S_2 will simultaneously have noncontextual determinate values, and *that* will run afoul of the theorems of Gleason and Kochen and Specker that were mentioned in Chapter 7. But maybe that worry (unlike the following two) need not be such a big one. States in which any two of the c_i's are *exactly* equal to one another, after all, will constitute a set of measure zero in the set of *all* of the possible quantum states of any given system. Perhaps there's a way of just ignoring them.

2. The KDH interpretations (as they've been written down so far) are radically *incomplete*. What's missing is a crucial chunk of the *dynamics*.

Here's what that means: As the quantum state of the world evolves, in accordance with the dynamical equations of motion, the observable of any given subsystem of the world that rule 1 picks out as the one that's presently *well-defined* is (of course) going to *change*. And the *question* that the missing chunk of the dynamics has to *answer* (which is analogous to the one that the *velocity algorithm* answers in *Bohm's* theory) is how the values of those observables that *have* values at t_2 are related to the values of those observables that *had* values at t_1.

And no workable idea of what that missing chunk of the dynamics could possibly look like has emerged yet.[2]

3. The measuring devices we actually *have* and *make use of* in the world are *never* absolutely "good" ones. In any *real* measurement, there is always some probability of the measuring device making an *error*. For any *real* measurement of (say) the hardness of an electron whose initial state is the one in (A.1), the post measurement state of the composite system consisting of m and e will invariably be something like

---

2. We do know, for example, that that chunk can't possibly be a deterministic one, as it is in Bohm's theory.

Here's how: Imagine a composite system which is initially prepared in a completely separable state (that is: a state in which there is a determinate matter of fact, on the standard way of thinking about what it means to be in a superposition, about the state of every individual *subsystem*). In a state like that, the observables that get picked out by rule 1 as well-defined, and the *values* of those observables, are (as the reader can easily confirm for herself) precisely the ones that get picked out by that state on the *standard* way of thinking.

And so knowing everything that there *is* to be known (on the KDH interpretations) about a system like that, when a quantum state like that obtains, is just a matter of knowing what that quantum state is.

And of course there isn't any matter of principle that stands in the way of anybody's knowing *that*.

And so, if that missing chunk of the dynamics were deterministic (which would mean that the entire dynamics of the world is deterministic), then anybody who merely knew that initial *quantum state* would be in a position to calculate, with certainty, what the values of all of those observables which *have* determinate values at any *future* time will be, even (for example) at times when the state of that composite system may have become *non*separable.

And that, of course, would violate rule 2, and it would violate our empirical experience of the world.

(A.4) $|J\rangle = c|\text{``hard''}\rangle_m|\text{hard}\rangle_e + d|\text{``soft''}\rangle_m|\text{soft}\rangle_e$
$\qquad + f|\text{``hard''}\rangle_m|\text{soft}\rangle_e + g|\text{``soft''}\rangle_m|\text{hard}\rangle_e$

in which the last two terms represent errors. In the more *expensive* sorts of hardness measuring devices, the coefficients f and g will be small; but the important point is that for *any* realizable such device, they will *invariably* be *nonzero*. And so long as f and g are nonzero, (A.4) is *not* a bi-orthogonal representation of the state $|J\rangle$.

Now, there will necessarily *be* some bi-orthogonal representation of $|J\rangle$. That representation will look something like this:

(A.5) $|J\rangle = k|w\rangle_m|@\rangle_e + l|z\rangle_m|@'\rangle_e$

where $|@\rangle$ and $|@'\rangle$ are eigenstates of some spin observable of e *other* than (and incompatible with) hardness, and $|w\rangle$ and $|z\rangle$ are eigenstates of some observable of m other than (and incompatible with) what m *indicates* about the hardness.

And so when $|J\rangle$ obtains rule 1 will *not* entail that the hardness of the electron has any determinate value, and (more importantly) it will *also* not entail that there is any determinate matter of fact about what m *indicates* the hardness of the electron to be.

And as a matter of fact, it turns out that no matter how small f and g are (that is: no matter how small the *imperfections of the measuring device* are), there will always be ways of choosing a and b in (A.1) such that the observable of e that gets picked out by rule 1 as well-defined, when $|J\rangle$ obtains, will be *maximally incompatible* with its hardness and (more importantly) such that the observable of m that gets picked out by rule 1 as well-defined, when $|J\rangle$ obtains, will be maximally incompatible with *what it indicates* about that hardness![3]

And (needless to say) whatever goes for interactions between macroscopic measuring devices and electrons will *also* go for inter-

3. It was Yakir Aharonov who first pointed this out to me. Here's how the argument goes:

Consider the following two states:

$|S1\rangle = a|\text{``hard''}\rangle_m|\text{hard}\rangle_e + b|\text{``soft''}\rangle_m|\text{soft}\rangle_e$
$|S2\rangle = a|\text{zip} = +1\rangle_m|\text{black}\rangle_e + b|\text{zip} = -1\rangle_m|\text{white}\rangle_e$

actions between sentient observers and macroscopic measuring devices.

And (consequently) if the world is (in any of the relevant senses) anything less than entirely perfect (and of course it invariably *is* something less than that), then the KDH interpretations don't end up doing their job right. And that's that.

There's an interesting variant on the KHD interpretations, which is due to Van Fraassen (Van Fraassen, 1991) and wherein the algorithm for picking out what's well-defined at any given moment depends not on the overall state of the world at that moment (which is what goes on in the KHD interpretations) but rather on what sorts of dynamical interactions led up to that state.

What Van Fraassen does is to present a general mathematical characterization of the sorts of dynamical interactions whereby

where

$$|\text{zip} = +1\rangle_m = \tfrac{1}{\sqrt{2}}(|\text{``hard''}\rangle_m + |\text{``soft''}\rangle_m)$$
$$|\text{zip} = -1\rangle_m = \tfrac{1}{\sqrt{2}}(|\text{``hard''}\rangle_m - |\text{``soft''}\rangle_m)$$

(and note that $|S1\rangle$ is the state in equation (A.2); the one that things end up in at the end of a good measurement of the hardness of an electron whose initial state is the one in (A.1)).

Now, if it happens to be the case that $a = -b$ (in the above expressions for $|S1\rangle$ and $|S2\rangle$), then (as the reader can easily confirm for herself, just by writing things out) $|S1\rangle$ and $|S2\rangle$ are precisely the same state; and (moreover) if it happens to be the case that a is *very nearly* equal to $-b$, *then* $|S1\rangle$ and $|S2\rangle$ are *almost* precisely the same state (that is: if a is very nearly equal to $-b$, then $\langle S1|S2\rangle$ is very nearly equal to 1).

And so in the event that a is sufficiently close to $-b$ in the initial state of the *measured electron* (that is: in the event that a is sufficiently close to $-b$ in the state in (A.1)), then an arbitrarily small imperfection in the measuring device (that is: an arbitrarily small *departure* from the initial conditions which represent a *perfect* measuring device, in its ready state) can have the effect of shifting the *post* measurement state of the *composite* system from $|S1\rangle$ to $|S2\rangle$.

And note that the observable of the electron that rule 1 picks out as well-defined in the state $|S2\rangle$ (which is its color) is maximally incompatible with its hardness, and note (and this is really the punch line) that the observable of the hardness measuring device that rule 1 picks out as well-defined in the state $|S2\rangle$ (which is its zip) is maximally incompatible with what the device *indicates* about hardness.

measurements can be accomplished (that is: what he does is to present a general mathematical characterization of the sorts of interactions whereby the value of some observable of some physical system after the interaction is over can in one way or another constitute a *record* of the value of some *other* observable of some *other* physical system before the interaction *began*).

And what his algorithm stipulates is that what's well-defined at the conclusion of any of those sorts of interactions is the value of the *record* observable.

But the trouble (here as before) is that none of the (imperfect) measurements which we actually *carry out* will ever precisely *satisfy* Van Fraassen's characterization. And so there isn't ever going to be a matter of fact (in the real world) about what observable the record observable *is*. And so Van Fraassen's algorithm will in general pick out *nothing whatsoever* (over and above what gets picked out by principles A–C of Chapter 2) as well-defined.

Bibliography

Aharonov, Y., and D. Bohm. 1959. Significance of electromagnetic potentials in the quantum theory. *Physical Review* 115:485.

Aicardi, F., A. Borsellino, G. C. Ghirardi, and R. Grassi. 1991. Dynamical models for state-vector reduction: do they ensure that measurements have outcomes? *Foundations of Physics Letters* 4:116.

Albert, D. 1983. On the theory of quantum-mechanical automata. *Physics Letters* 98A:249–252.

―――― 1988. On the possibility that the present quantum state of the universe is the vacuum. *Proceedings of the 1988 Biennial Meeting of the Philosophy of Science Association,* vol. 2, pp. 127–133, ed. A. Fine and J. Leplin. East Lansing: Philosophy of Science Association.

Albert, D., and B. Loewer. 1988. Interpreting the many-worlds interpretation. *Synthese* 77:195–213.

―――― 1991. The measurement problem: some "solutions." *Synthese* 86:87–98.

Albert, D., and L. Viadman. 1988. On a proposed postulate of state-reduction. *Physics Letters* 139A:1–4.

Barrett, J. 1992. Doctoral thesis, Department of Philosophy, Columbia University (in preparation).

Bell, J. 1964. On the Einstein-Podolsky-Rosen paradox. *Physics* 1:195–200. (Reprinted in Bell, 1987b.)

―――― 1976. How to teach special relativity. *Progress in Scientific Culture,* vol. 1. (Reprinted in Bell, 1987b.)

―――― 1982. On the impossible pilot wave. *Foundations of Physics* 12:989–999. (Reprinted in Bell, 1987b.)

―――― 1984. Beables for quantum field theory. *CERN-TH* 4035/84. (Reprinted in Bell, 1987b.)

―――― 1987a. Are there quantum jumps? In *Schrödinger: Centenary of a Polymath.* Cambridge: Cambridge University Press (Reprinted in Bell, 1987b.)

―――― 1987b. *Speakable and Unspeakable in Quantum Mechanics.* Cambridge: Cambridge University Press.

Bohm, D. 1952. A suggested interpretation of quantum theory in terms of "hidden variables," parts I and II. *Physical Review* 85:166–193. (Reprinted in Wheeler and Zurek, 1983.)

Daneri, A., A. Loinger, and G. M. Prosperi. 1962. Quantum theory of measurement and ergoticity conditions. *Nuclear Physics* 33:297–319. (Reprinted in Wheeler and Zurek, 1983.)

de Broglie, L. 1930. *Rapport au V'ieme Congres de Physique Solvay.* Paris: Gauther-Villars.

d'Espagnat, B. 1971. *Conceptual Foundations of Quantum Mechanics.* Menlo Park: W. A. Benjamin.

DeWitt, B. 1970. Quantum mechanics and reality. *Physics Today* 23:30–35.

Dieks, D. 1991. On some alleged difficulties in interpretation of quantum mechanics. *Synthese* 86:77–86.

Einstein, A., B. Podolsky, and N. Rosen. 1935. Can the quantum-mechanical description of physical reality be considered complete? *Physical Review* 47:777–780. (Reprinted in Wheeler and Zurek, 1983.)

Everett, H., III. 1957. Relative state formulation of quantum mechanics. *Reviews of Modern Physics* 29:454–462. (Reprinted in Wheeler and Zurek, 1983.)

Gell-Mann, M., and J. B. Hartle. 1990. Quantum mechanics in the light of quantum cosmology. In *Complexity, Entropy, and the Physics of Information,* Santa Fe Institute Studies in the Sciences of Complexity, vol. 8, ed. W. Zurek. Reading: Addison-Wesley.

Ghirardi, G. C., A. Rimini, and T. Weber. 1986. Unified dynamics for microscopic and macroscopic systems. *Physical Review* D34:470.

Gleason, A. M. 1957. Measures on the closed subspaces of a Hilbert space. *Journal of Mathematics and Mechanics* 6:885–893.

Gottfried, K. 1966. *Quantum Mechanics,* vol. 1. Reading: W. A. Benjamin.

Healy, R. 1989. *The Philosophy of Quantum Mechanics.* New York: Cambridge University Press.

Kochen, S. 1985. A new interpretation of quantum mechanics. In *Symposium on the Foundations of Modern Physics,* ed. P. Lahti and P. Mittelstaedt, pp. 151–170. Singapore: World Scientific.

Kochen, S., and E. P. Specker. 1967. The problem of hidden variables in quantum mechanics. *Journal of Mathematics and Mechanics* 17:59–87.

Lockwood, M. 1989. *Mind, Brain, and the Quantum.* Oxford: Basil Blackwell.

Peres, A. 1980. Can we undo quantum measurements? *Physical Review* D22:879–883. (Reprinted in Wheeler and Zurek, 1983.)

Putnam, H., and D. Albert. 1992. Further adventures of Schrödinger's cat. Unpublished manuscript.

Van Fraassen, B. 1991. *Quantum Mechanics: An Empiricist View.* Oxford: Oxford University Press.

Von Neumann, J. 1955. *Mathematical Foundations of Quantum Mechanics.* Trans. R. T. Beyer. Princeton: Princeton University Press. (First published in German in 1932.)

Wheeler, J., and W. Zurek, eds. 1983. *Quantum Theory and Measurement.* Princeton: Princeton University Press.

Wigner, E. P. 1961. Remarks on the mind-body question. In *The Scientist Speculates,* ed. I. J. Good, pp. 284–302. New York: Basic Books. (Reprinted in Wheeler and Zurek, 1983.)

Index